High Performance Datacenter Networks

Architectures, Algorithms, and Opportunities

Synthesis Lectures on Computer Architecture

Editor
Mark D. Hill, *University of Wisconsin*

Synthesis Lectures on Computer Architecture publishes 50- to 100-page publications on topics pertaining to the science and art of designing, analyzing, selecting and interconnecting hardware components to create computers that meet functional, performance and cost goals. The scope will largely follow the purview of premier computer architecture conferences, such as ISCA, HPCA, MICRO, and ASPLOS.

High Performance Datacenter Networks: Architectures, Algorithms, and Opportunities
Dennis Abts and John Kim
2011

Quantum Computing for Architects, Second Edition
Tzvetan Metodi, Fred Chong, and Arvin Faruque
2011

Processor Microarchitecture: An Implementation Perspective
Antonio González, Fernando Latorre, and Grigorios Magklis
2010

Transactional Memory, 2nd edition
Tim Harris, James Larus, and Ravi Rajwar
2010

Computer Architecture Performance Evaluation Methods
Lieven Eeckhout
2010

Introduction to Reconfigurable Supercomputing
Marco Lanzagorta, Stephen Bique, and Robert Rosenberg
2009

On-Chip Networks
Natalie Enright Jerger and Li-Shiuan Peh
2009

High Performance Datacenter Networks: Architectures, Algorithms, and Opportunities

Dennis Abts and John Kim

ISBN: 978-3-031-00602-9   paperback
ISBN: 978-3-031-01730-8   ebook

DOI 10.1007/978-3-031-01730-8

A Publication in the Springer series
SYNTHESIS LECTURES ON COMPUTER ARCHITECTURE

Lecture #14
Series Editor: Mark D. Hill, *University of Wisconsin*
Series ISSN
Synthesis Lectures on Computer Architecture
Print 1935-3235   Electronic 1935-3243

High Performance Datacenter Networks

Architectures, Algorithms, and Opportunities

Dennis Abts
Google Inc.

John Kim
Korea Advanced Institute of Science and Technology (KAIST)

SYNTHESIS LECTURES ON COMPUTER ARCHITECTURE #14

ABSTRACT

Datacenter networks provide the communication substrate for large parallel computer systems that form the ecosystem for high performance computing (HPC) systems and modern Internet applications. The design of new datacenter networks is motivated by an array of applications ranging from communication intensive climatology, complex material simulations and molecular dynamics to such Internet applications as Web search, language translation, collaborative Internet applications, streaming video and voice-over-IP. For both Supercomputing and Cloud Computing the network enables distributed applications to communicate and interoperate in an orchestrated and efficient way.

    This book describes the design and engineering tradeoffs of datacenter networks. It describes interconnection networks from topology and network architecture to routing algorithms, and presents opportunities for taking advantage of the emerging technology trends that are influencing router microarchitecture. With the emergence of "many-core" processor chips, it is evident that we will also need "many-port" routing chips to provide a bandwidth-rich network to avoid the performance limiting effects of Amdahl's Law. We provide an overview of conventional topologies and their routing algorithms and show how technology, signaling rates and cost-effective optics are motivating new network topologies that scale up to millions of hosts. The book also provides detailed case studies of two high performance parallel computer systems and their networks.

KEYWORDS

network architecture and design, topology, interconnection networks, fiber optics, parallel computer architecture, system design

Contents

Preface

This book is aimed at the researcher, graduate student and practitioner alike. We provide some background and motivation to provide the reader with a substrate upon which we can build the new concepts that are driving high-performance networking in both supercomputing and cloud computing. We assume the reader is familiar with computer architecture and basic networking concepts. We show the evolution of high-performance interconnection networks over the span of two decades, and the underlying technology trends driving these changes. We describe how to apply these technology drivers to enable new network topologies and routing algorithms that scale to millions of processing cores. We hope that practitioners will find the material useful for making design tradeoffs, and researchers will find the material both timely and relevant to modern parallel computer systems which make up today's datacenters.

Dennis Abts and John Kim
March 2011

Acknowledgments

While we draw from our experience at Cray and Google and academic work on the design and operation of interconnection networks, most of what we learned is the result of hard work, and years of experience that have led to practical insights. Our experience benefited tremendously from our colleagues Steve Scott at Cray, and Bill Dally at Stanford University. In addition, many hours of whiteboard-huddled conversations with Mike Marty, Philip Wells, Hong Liu, and Peter Klausler at Google. We would also like to thank Google colleagues James Laudon, Bob Felderman, Luiz Barroso, and Urs Hölzle for reviewing draft versions of the manuscript. We want to thank the reviewers, especially Amin Vahdat and Mark Hill for taking the time to carefully read and provide feedback on early versions of this manuscript. Thanks to Urs Hölzle for guidance, and Kristin Weissman at Google and Michael Morgan at Morgan & Claypool Publishers. Finally, we are grateful for Mark Hill and Michael Morgan for inviting us to this project and being patient with deadlines.

Finally, and most importantly, we would like to thank our loving family members who graciously supported this work and patiently allowed us to spend our free time to work on this project. Without their enduring patience and with an equal amount of prodding, this work would not have materialized.

Dennis Abts and John Kim
March 2011

Note to the Reader

We very much appreciate any feedback, suggestions, and corrections you might have on our manuscript. The Morgan & Claypool publishing process allows a lightweight method to revise the electronic edition. We plan to revise the manuscript relatively often, and will gratefully acknowledge any input that will help us to improve the accuracy, readability, or general usefulness of the book. Please leave your feedback at http://tinyurl.com/HPNFeedback

Dennis Abts and John Kim
March 2011

CHAPTER 1

Introduction

Today's datacenters have emerged from the collection of loosely connected workstations, which shaped the humble beginnings of the Internet, and grown into massive "warehouse-scale computers" (Figure 1.1) capable of running the most demanding workloads. Barroso and Hölzle describe the architecture of a warehouse-scale computer (WSC) [9] and give an overview of the programming model and common workloads executed on these machines. The hardware building blocks are packaged into "racks" of about 40 servers, and many racks are interconnected using a high-performance network to form a "cluster" with hundreds or thousands of *tightly-coupled* servers for performance,

Figure 1.1: A datacenter with cooling infrastructure and power delivery highlighted.

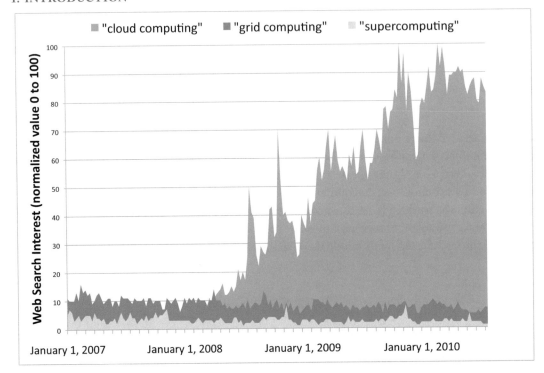

Figure 1.2: Comparison of web search interest and terminology.

but *loosely-coupled* for fault tolerance and isolation. This highlights some distinctions between what have traditionally been called "supercomputers" and what we now consider "cloud computing," which appears to have emerged around 2008 (based on the relative Web Search interest shown in Figure 1.2) as a moniker for *server-side* computing. Increasingly, our computing needs are moving away from desktop computers toward more mobile clients (e.g., smart phones, tablet computers, and net-books) that depend on Internet services, applications, and storage. As an example, it is much more efficient to maintain a repository of digital photography on a server in the "cloud" than on a PC-like computer that is perhaps not as well maintained as a server in a large datacenter, which is more reminiscent of a clean room environment than a living room where your precious digital memories are subjected to the daily routine of kids, spills, power failures, and varying temperatures; in addition, most consumers upgrade computers every few years, requiring them to migrate *all* their precious data to their newest piece of technology. In contrast, the "cloud" provides a clean, temperature controlled environment with ample power distribution and backup. Not to mention your data in the "cloud" is probably replicated for redundancy in the event of a hardware failure the user data is replicated and restored generally without the user even aware that an error occurred.

1.1   FROM SUPERCOMPUTING TO CLOUD COMPUTING

As the ARPANET transformed into the Internet over the past forty years, and the World Wide Web emerges from adolescence and turns twenty, this metamorphosis has seen changes in both supercomputing and cloud computing. The supercomputing industry was born in 1976 when Seymour Cray announced the Cray-1 [54]. Among the many innovations were its processor design, process technology, system packaging, and instruction set architecture. The foundation of the architecture was based on the notion of *vector* operations that allowed a single instruction to operate on an array, or "vector," of elements simultaneously. In contrast to *scalar* processors of the time whose instructions operated on single data items. The *vector* parallelism approach dominated the high-performance computing landscape for much of the 1980s and early 1990s until "commodity" microprocessors began aggressively implementing forms of instruction-level parallelism (ILP) and better cache memory systems to exploit spatial and temporal locality exhibited by most applications. Improvements in CMOS process technology and full-custom CMOS design practices allowed microprocessors to quickly ramp up clock rates to several gigahertz. This coupled with multi-issue pipelines; efficient branch prediction and speculation eventually allowed microprocessors to catch up with their proprietary vector processors from Cray, Convex, and NEC. Over time, conventional microprocessors incorporated short vector units (e.g., SSE, MMX, AltiVec) into the instruction set. However, the largest beneficiary of vector processing has been multimedia applications as evidenced by the jointly developed (by Sony, Toshiba, and IBM) Cell processor which found widespread success in Sony's Playstation3 game console, and even some special-purpose computer systems like Mercury Systems.

Parallel applications eventually have to synchronize and communicate among parallel threads. Amdahl's Law is relentless and unless enough parallelism is exposed, the time spent orchestrating the parallelism and executing the sequential region will ultimately limit the application performance [27].

1.2   BEOWULF: THE CLUSTER IS BORN

In 1994 Thomas Sterling (then dually affiliated with the California Institute of Technology and NASAs JPL) and Donald Becker (then a researcher at NASA) assembled a parallel computer that became known as a *Beowulf* cluster[1]. What was unique about Beowulf [61] systems was that they were built from common "off-the-shelf" computers, as Figure 1.3 shows, system packaging was not an emphasis. More importantly, as a loosely-coupled distributed memory machine, Beowulf forced researchers to think about *how* to efficiently program parallel computers. As a result, we benefited from portable and free programming interfaces such as parallel virtual machines (PVM), message passing interfaces (MPICH and OpenMPI), local area multiprocessor (LAM); with MPI being embraced by the HPC community and highly optimized.

The Beowulf cluster was organized so that one machine was designated the "server," and it managed job scheduling, pushing binaries to clients, and monitoring. It also acted as the gateway

---

[1]The genesis of the name comes from the poem which describes Beowulf as having "thirty men's heft of grasp in the gripe of his hand."

Figure 1.3: An 128 processor Beowulf cluster at NASA.

to the "outside world," so researchers had a login host. The model is still quite common: with some nodes being designated as service and IO nodes where users actually login to the parallel machine. From there, they can compile their code, and launch the job on "compute only" nodes — the worker bees of the colony — and console information, machine status is communicated to the service nodes.

1.3   OVERVIEW OF PARALLEL PROGRAMMING MODELS

Early supercomputers were able to work efficiently, in part, because they shared a common *physical* memory space. As a result, communication among processors was very efficient as they updated shared variables and operated on common data. However, as the size of the systems grew, this shared memory model evolved into a *distributed shared memory* (DSM) model where each processing node owns a portion of the machines physical memory and the programmer is provided with a *logically shared* address space making it easy to reason about how the application is partitioned and communication among threads. The Stanford DASH [45] was the first to demonstrate this cache-coherent non-uniform memory (ccNUMA) access model, and the SGI Origin2000 [43] was the first machine to successfully commercialize the DSM architecture.

We commonly refer to *distributed memory* machines as "clusters" since they are loosely-coupled and rely on message passing for communication among processing nodes. With the inception of Beowulf clusters, the HPC community realized they could build modest-sized parallel computers on a relatively small budget. To their benefit, the common benchmark for measuring the performance of a parallel computer is LINPACK, which is not communication intensive, so it was commonplace to use inexpensive Ethernet networks to string together commodity nodes. As a result, Ethernet got a foothold on the list of the TOP500 [62] civilian supercomputers with almost 50% of the TOP500 systems using Ethernet.

1.4   PUTTING IT ALL TOGETHER

The first Cray-1 [54] supercomputer had expected to ship one system per quarter in 1977. Today, microprocessor companies have refined their CMOS processes and manufacturing making them very cost-effective building blocks for large-scale parallel systems capable of 10s of petaflops. This shift away from "proprietary" processors and trend toward "commodity" processors has fueled the growth of systems. At the time of this writing, the largest computer on the TOP500 list [62] has in excess of 220,000 cores (see Figure 7.5) and consumes almost seven megawatts!

A datacenter server has many commonalities as one used in a supercomputer, however, there are also some very glaring differences. We enumerate several properties of both a warehouse-scale computer (WSC) and a supercomputer (Cray XE6).

Datacenter server

- **Sockets per server** 2 sockets x86 platform

- **Memory capacity** 16 GB DRAM

- **Disk capacity** 5×1TB disk drive, and 1×160GB SSD (FLASH)

- **Compute density** 80 sockets per rack

- **Network bandwidth per rack** 1×48-port GigE switch with 40 down links, and 8 uplinks ($5 \times$ oversubscription)

- **Network bandwidth per socket** 100 Mb/s if 1 GigE rack switch, or 1 Gb/s if 10 GigE rack switch

Supercomputer server

- **Sockets per server** 8 sockets x86 platform

- **Memory capacity** 32 or 64 GB DRAM

- **Disk capacity** IO capacity varies. Each XIO blade has four PCIe-Gen2 interfaces, for a total of 96 PCIe-Gen2 $\times 16$ IO devices for a peak IO bandwidth of 768 GB/s per direction.

- **Compute density** 192 sockets per rack

- **Network bandwidth per rack** 48×48-port Gemini switch chips each with 160 GB/s switching bandwidth

- **Network bandwidth per socket** 9.6GB/s injection bandwidth with non-coherent Hyper-Transport 3.0 (ncHT3)

Several things stand out as differences between a datacenter server and supercomputer node. First, the *compute density* for the supercomputer is significantly better than a standard 40U rack. On the other hand, this dense packaging also puts pressure on cooling requirements not to mention power delivery. As power and its associated delivery become increasingly expensive, it becomes more important to optimize the number of operations per watt; often the *size* of a system is limited by power distribution and cooling infrastructure.

Another point is the vast difference in *network bandwidth per socket* in large part because ncHT3 is a much higher bandwidth processor interface than PCIe-Gen2, however, as PCI-Gen3×16 becomes available we expect that gap to narrow.

1.5   QUALITY OF SERVICE (QOS) REQUIREMENTS

With HPC systems it is commonplace to dedicate the system for the duration of application execution. Allowing *all* processors to be used for compute resources. As a result, there is no need for *performance isolation* from competing applications. *Quality of Service (QoS)* provides both performance isolation *and differentiated service* for applications[2]. Cloud computing often has a varied workloads requiring multiple applications to share resources. *Workload consolidation* [33] is becoming increasingly important as memory and processor cost increase, as a result so does the value of increased system utilization.

The QoS *class* refers to the end-to-end class of service as observed by the application. In principle, QoS is divided into three categories:

Best effort - traffic is treated as a FIFO with no differentiation provided.

Differentiated service - also referred to as "soft QoS" where traffic is given a statistical preference over other traffic. This means it is less likely to be dropped relative to best effort traffic, for example, resulting in lower average latency and increased average bandwidth.

Guaranteed service - also referred to as "hard QoS" where a fraction of the network bandwidth is reserved to provide no-loss, low jitter bandwidth guarantees.

In practice, there are many intermediate pieces which are, in part, responsible for implementing a QoS scheme. A routing algorithm determines the set of usable paths through the network between any source and destination. Generally speaking, routing is a background process that attempts to load-balance the physical links in the system taking into account any network faults and programming

---

[2]We use the term "applications" loosely here to represent processes or threads, at whatever granularity a service level agreement is applied.

the forwarding tables within each router. When a new packet arrives, the header is inspected and the network address of the destination is used to index into the forwarding table which emits the output port where the packet is scheduled for transmission. The "packet forwarding" process is done on a packet-by-packet basis and is responsible for identifying packets marked for special treatment according to its QoS class.

The basic unit over which a QoS class is applied is the flow. A flow is described as a tuple (SourceIP, SourcePort, DestIP, DestPort). Packets are marked by the host or edge switch using either 1) port range, or 2) host (sender/client-side) marking. Since we are talking about end-to-end service levels, ideally the host which initiates the communication would request a specific level of service. This requires some client-side API for establishing the QoS requirements prior to sending a message. Alternatively, edge routers can mark packets as they are injected into the core fabric.

Packets are marked with their service class which is interpreted at each hop and acted upon by routers along the path. For common Internet protocols, the differentiated service (DS) field of the IP header provides this function as defined by the DiffServ [RFC2475] architecture for network layer QoS. For compatibility reasons, this is the same field as the *type of service* (ToS) field [RFC791] of the IP header. Since the RFC does not clearly describe how "low," "medium," or "high" are supposed to be interpreted, it is common to use five classes: best effort (BE), AF1, AF2, AF3, AF4, and set the drop priority to 0 (ignored).

1.6 FLOW CONTROL

Surprisingly, a key difference in system interconnects is *flow control*. How the switch and buffer resources are managed is very different in Ethernet than what is typical in a supercomputer interconnect. There are several kinds of flow control in a large distributed parallel computer. The interconnection network is a shared resource among all the compute nodes, and network resources must be carefully managed to avoid corrupting data, overflowing a buffer, etc. The basic mechanism by which resources in the network are managed is *flow control*. Flow control provides a simple accounting method for managing resources that are in demand by multiple uncoordinated sources. The resource is managed in units of *flits* (flow control units). When a resource is requested but not currently available for use, we must decide what to do with the incoming request. In general, we can 1) drop the request and all subsequent requests until the resource is freed, or 2) block and wait for the request to free.

1.6.1 LOSSY FLOW CONTROL

With a lossy flow control [20, 48], the hardware can discard packets until there is room in the desired resource. This approach is usually applied to *input buffers* on each switch chip, but also applies to resources in the network interface controller (NIC) chip as well. When packets are dropped, the software layers must detect the loss, usually through an unexpected *sequence number* indicating that one or more packets are missing or out of order. The receiver software layers will discard packets that do not match the expected sequence number, and the sender software layers will detect that it

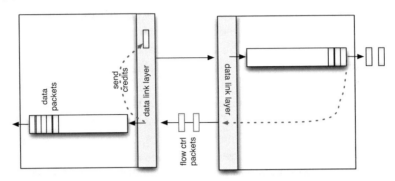

Figure 1.4: Example of credit-based flow control across a network link.

has not received an *acknowledgment* packet and will cause a sender timeout which prompts the "send window" — packets sent since the last acknowledgment was received — to be retransmitted. This algorithm is referred to as go-back-N since the sender will "go back" and retransmit the last N (send window) packets.

1.6.2   LOSSLESS FLOW CONTROL

Lossless flow control implies that packets are never dropped as a results of lack of buffer space (i.e., in the presence of congestion). Instead, it provides *back pressure* to indicate the absence of available buffer space in the resource being managed.

1.6.2.1   Stop/Go (XON/XOFF) flow control

A common approach is *XON/XOFF* or *stop/go* flow control. In this approach, the receiver provides simple handshaking to the sender indicating whether it is safe (XON) to transmit, or not (XOFF). The sender is able to send flits until the receiver asserts *stop* (XOFF). Then, as the receiver continues to process packets from the input buffer freeing space, and when a *threshold* is reached the receiver will assert the XON again allowing the sender to again start sending. This Stop/Go functionality correctly manages the resource and avoids overflow as long as the time at which XON is asserted again (i.e., the threshold level in the input buffer) minus the time XOFF is asserted and the buffer is sufficient to allow any in-flight flits to land. This *slack* in the buffer is necessary to act as a flow control shock absorber for outstanding flits necessary to cover the propagation delay of the flow control signals.

1.6.2.2   Credit-based flow control

Credit based flow control (Figure 1.4) provides more efficient use of the buffer resources. The sender maintains a count of the number of available *credits*, which represent the amount of free space in the receiver's input buffer. A separate count is used for each virtual channel (VC) [21]. When a new packet arrives at the output port, the sender checks the available credit counter. For *wormhole* flow control [20] across the link, the sender's available credit needs to only be one or more. For *virtual cut-through* (VCT) [20, 22] flow control across the link, the sender's available credit must be more than the size of the packet. In practice, the switch hardware doesn't have to track the *size* of the packet in order to allow VCT flow control. The sender can simply check the available credit count is larger than the maximum packet size.

1.7   THE RISE OF ETHERNET

It may be an extreme example comparing a typical datacenter server to a state-of-the-art supercomputer node, but the fact remains that Ethernet is gaining a significant foothold in the high-performance computing space with nearly 50% of the systems on the TOP500 list [62] using Gigabit Ethernet as shown in Figure 1.5(b). Infiniband (includes SDR, DDR and QDR) accounts for 41% of the interconnects leaving very little room for proprietary networks. The landscape was very different in 2002, as shown in Figure 1.5(a), where Myrinet accounted for about one third of the system interconnects. The IBM SP2 interconnect accounted for about 18%, and the remaining 50% of the system interconnects were split among about nine different manufacturers. In 2002, only about 8% of the TOP500 systems used gigabit Ethernet, compared to the nearly 50% in June of 2010.

1.8   SUMMARY

No doubt "cloud computing" benefited from this wild growth and acceptance in the HPC community, driving prices down and making more reliable parts. Moving forward we may see even further consolidation as 40 Gig Ethernet converges with some of the Infiniband semantics with RDMA over Ethernet (ROE). However, a warehouse-scale computer (WSC) [9] and a supercomputer have different usage models. For example, most supercomputer applications expect to run on the machine in a dedicated mode, not having to compete for compute, network, or IO resources with *any* other applications.

Supercomputing applications will commonly checkpoint their dataset, since the MTBF of a large system is usually measured in 10s of hours. Supercomputing applications also typically run with a dedicated system, so QoS demands are not typically a concern. On the other hand, a datacenter will run a wide variety of applications, some user-facing like Internet email, and others behind the scenes. The workloads vary drastically, and programmers must learn that hardware can, and does, fail and the application must be fault-aware and deal with it gracefully. Furthermore, clusters in the datacenter are often shared across dozens of applications, so performance isolation and fault isolation are key to scaling applications to large processor counts.

Choosing the "right" topology is important to the overall system performance. We must take into account the flow control, QoS requirements, fault tolerance and resilience, as well as workloads to better understand the latency and bandwidth characteristics of the entire system. For example,

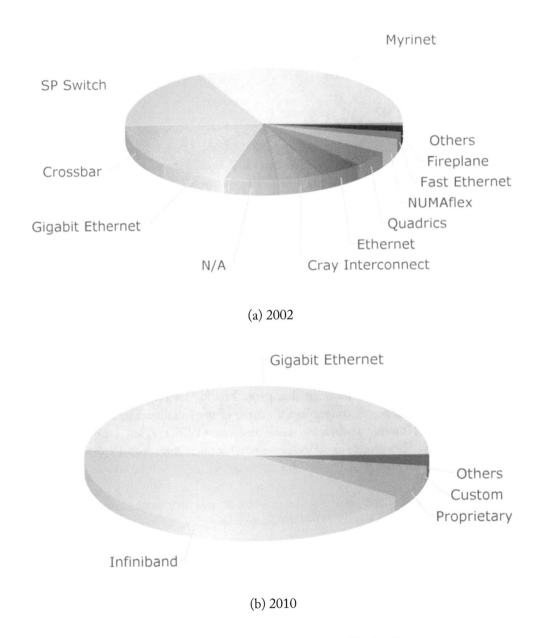

(a) 2002

(b) 2010

Figure 1.5: Breakdown of supercomputer interconnects from the Top500 list.

topologies with abundant *path diversity* are able to find alternate routes between arbitrary endpoints. This is only one aspect of topology choice that we will consider in subsequent chapters.

CHAPTER 2

Background

Over the past three decades, Moore's Law has ushered in an era where transistors within a single silicon package are abundant; a trend that system architects took advantage of to create a class of many-core chip multiprocessors (CMPs) which interconnect many small processing cores using an on-chip network. However, the *pin density*, or number of signal pins per unit of silicon area, has not kept up with this pace. As a result *pin bandwidth*, the amount of data we can get on and off the chip package, has become a first-order design constraint and precious resource for system designers.

2.1 INTERCONNECTION NETWORKS

The components of a computer system often have to communicate to exchange status information, or data that is used for computation. The *interconnection network* is the substrate over which this communication takes place. Many-core CMPs employ an *on-chip* network for low-latency, high-bandwidth load/store operations between processing cores and memory, and among processing cores within a chip package.

Processor, memory, and its associated IO devices are often packaged together and referred to as a *processing node*. The *system-level* interconnection network connects all the processing nodes according to the network *topology*. In the past, system components shared a *bus* over which address and data were exchanged, however, this communication model did not scale as the number of components sharing the bus increased. Modern interconnection networks take advantage of high-speed signaling [28] with *point-to-point* serial links providing high-bandwidth connections between processors and memory in multiprocessors [29, 32], connecting input/output (IO) devices [31, 51], and as switching fabrics for routers.

2.2 TECHNOLOGY TRENDS

There are many considerations that go into building a large-scale cluster computer, many of which revolve around its cost effectiveness, in both *capital* (procurement) cost and *operating* expense. Although many of the components that go into a cluster each have different *technology* drivers which blurs the line that defines the optimal solution for both performance and cost. This chapter takes a look at a few of the technology drivers and how they pertain to the interconnection network.

The interconnection network is the substrate over which processors, memory and I/O devices interoperate. The underlying technology from which the network is built determines the data rate, resiliency, and cost of the network. Ideally, the processor, network, and I/O devices are all orchestrated in a way that leads to a cost-effective, high-performance computer system. The system, however, is no better than the components from which it is built.

The basic building block of the network is the *switch* (router) chip that interconnects the processing nodes according to some prescribed *topology*. The topology and how the system is packaged are closely related; typical packaging schemes are hierarchical – chips are packaged onto printed circuit boards, which in turn are packaged into an enclosure (e.g., rack), which are connected together to create a single system.

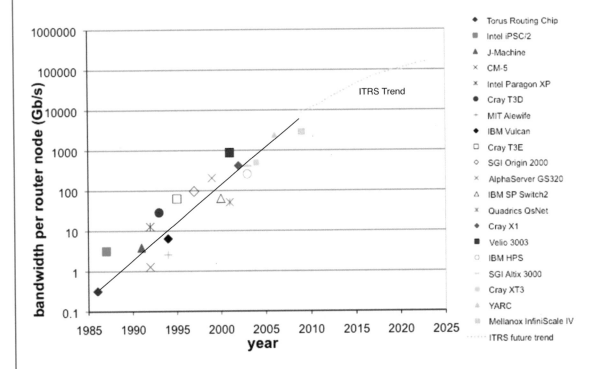

Figure 2.1: Off-chip bandwidth of prior routers, and ITRS predicted growth.

The past 20 years has seen several orders of magnitude increase in off-chip bandwidth spanning from several gigabits per second up to several terabits per second today. The bandwidth shown in Figure 2.1 plots the total pin bandwidth of a router – i.e., equivalent to the total number of signals times the signaling rate of each signal – and illustrates an exponential increase in pin bandwidth. Moreover, we expect this trend to continue into the next decade as shown by the International Roadmap for Semiconductors (ITRS) in Figure 2.1, with 1000s of pins per package and more than 100 Tb/s of off-chip bandwidth. Despite this exponential growth, pin and wire density simply does not match the growth rates of transistors as predicted by Moore's Law.

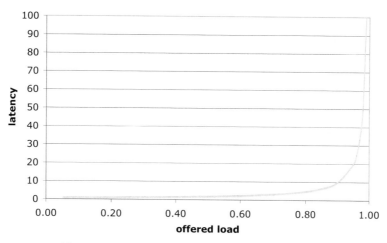

(a) Load versus latency for an ideal M/D/1 queue model.

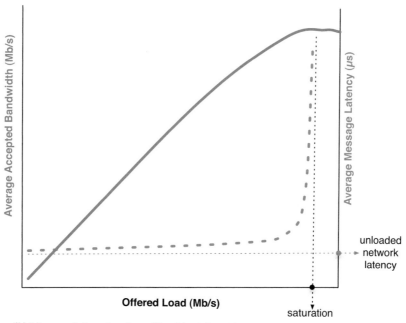

(b) Measured data showing offered load (Mb/s) versus latency (μs) with average accepted throughput (Mb/s) overlaid to demonstrate saturation in a real network.

Figure 2.2: Network latency and bandwidth characteristics.

2.3   TOPOLOGY, ROUTING AND FLOW CONTROL

Before diving into details of what drives network performance, we pause to lay the ground work for some fundamental terminology and concepts. Network performance is characterized by its latency and bandwidth characteristics as illustrated in Figure 2.2. The queueing delay, $Q(\lambda)$, is a function of the offered load (λ) and described by the latency-bandwidth characteristics of the network. An approximation of $Q(\lambda)$ is given by an M/D/1 queue model, Figure 2.2(a). If we overlay the average accepted bandwidth observed by each node, assuming benign traffic, we Figure 2.2(b).

$$Q(\lambda) = \frac{1}{1 - \lambda} \tag{2.1}$$

When there is very low offered load on the network, the $Q(\lambda)$ delay is negligible. However, as traffic intensity increases, and the network approaches saturation, the queueing delay will dominate the total packet latency.

The performance and cost of the interconnect are driven by a number of design factors, including topology, routing, flow control, and message efficiency. The *topology* describes how network nodes are interconnected and determines the *path diversity* — the number of distinct paths between any two nodes. The *routing algorithm* determines which path a packet will take in such as way as to *load balance* the physical links in the network. Network resources (primarily buffers for packet storage) are managed using a *flow control* mechanism. In general, flow control happens at the link-layer and possibly end-to-end. Finally, packets carry a *data payload* and the *packet efficiency* determines the *delivered* bandwidth to the application.

While recent *many-core* processors have spurred a $2\times$ and $4\times$ increase in the number of processing cores in each cluster, unless network performance keeps pace, the effects of Amdahl's Law will become a limitation. The topology, routing, flow control, and message efficiency all have first-order affects on the system performance, thus we will dive into each of these areas in more detail in subsequent chapters.

2.4   COMMUNICATION STACK

Layers of abstraction are commonly used in networking to provide fault isolation and device independence. Figure 2.3 shows the communication stack that is largely representative of the lower four layers of the OSI networking model. To reduce software overhead and the resulting end-to-end latency, we want a thin networking stack. Some of the protocol processing that is common in Internet communication protocols is handled in specialized hardware in the network interface controller (NIC). For example, the *transport* layer provides reliable message delivery to applications and whether the protocol bookkeeping is done in software (e.g., TCP) or hardware (e.g., Infiniband reliable connection) directly affects the application performance. The *network* layer provides a logical namespace for endpoints (and possibly switches) in the system. The network layer handles *packets*, and provides the *routing* information identifying paths through the network among all source, destination pairs. It is the network layer that asserts routes, either at the source (i.e., source-routed)

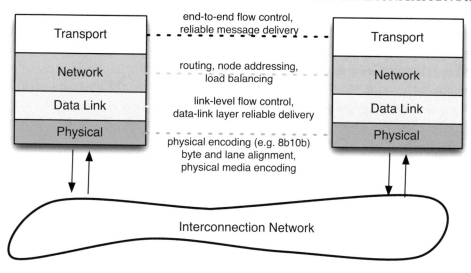

Figure 2.3: The communication stack.

or along each individual hop (i.e., distributed routing) along the path. The *data link* layer provides link-level flow control to manage the receiver's input buffer in units of *flits* (flow control units). The lowest level of the protocol stack, the *physical media* layer, is where data is encoded and driven onto the medium. The physical encoding must maintain a DC-neutral transmission line and commonly uses 8b10b or 64b66b encoding to balance the transition density. For example, a 10-bit encoded value is used to represent 8-bits of data resulting in a 20% physical encoding overhead.

SUMMARY

Interconnection networks are a critical component of modern computer systems. The emergence of *cloud* computing, which provides a homogenous cluster using conventional microprocessors and common Internet communication protocols aimed at providing Internet services (e.g., email, Web search, collaborative Internet applications, streaming video, and so forth) at large scale. While Internet services themselves may be insensitive to latency, since they operate on human timescales measured in 100s of milliseconds, the backend applications providing those services may indeed require large amounts of bandwidth (e.g., indexing the Web) and low latency characteristics. The programming model for cloud services is built largely around distributed message passing, commonly implemented around TCP (transport control protocol) as a conduit for making a remote procedure call (RPC).

Supercomputing applications, on the other hand, are often communication intensive and can be sensitive to network latency. The programming model may use a combination of shared memory and message passing (e.g., MPI) with often very fine-grained communication and synchronization needs. For example, *collective* operations, such as global sum, are commonplace in supercomputing applications and rare in Internet services. This is largely because Internet applications evolved from simple hardware primitives (e.g., low-cost ethernet NIC) and common communication models (e.g., TCP sockets) that were incapable of such operations.

As processor and memory performance continues to increase, the interconnection network is becoming increasingly important and largely determines the bandwidth and latency of remote memory access. Going forward, the emergence of *super datacenters* will convolve into *exa*-scale parallel computers.

CHAPTER 3
Topology Basics

The network *topology* — describing precisely how nodes are connected — plays a central role in both the performance and cost of the network. In addition, the topology drives aspects of the switch design (e.g., virtual channel requirements, routing function, etc), fault tolerance, and sensitivity to adversarial traffic. There are subtle yet very practical design issues that only arise *at scale*; we try to highlight those key points as they appear.

3.1 INTRODUCTION

Many scientific problems can be decomposed into a 3-D structure that represents the basic building blocks of the underlying phenomenon being studied. Such problems often have nearest neighbor communication patterns, for example, and lend themselves nicely to k-ary n-cube networks. A high-performance application will often use the system *dedicated* to provide the necessary performance isolation, however, a large production datacenter cluster will often run multiple applications simultaneously with varying workloads and often unstructured communication patterns.

The choice of topology is largely driven by two factors: technology and packaging constraints. Here, *technology* refers to the underlying silicon from which the routers are fabricated (i.e., node size, pin density, power, etc) and the signaling technology (e.g., optical versus electrical). The packaging constraints will determine the *compute density*, or amount of computation per unit of area on the datacenter floor. The packaging constraints will also dictate the data rate (signaling speed) and distance over which we can reliably communicate.

As a result of evolving technology, the topologies used in large-scale systems have also changed. Many of the earliest interconnection networks were designed using topologies such as butterflies or hypercubes, based on the simple observation that these topologies minimized hop count. Analysis by both Dally [18] and Agarwal [5] showed that under fixed packaging constraints, a *low-radix* network offered lower packet latency and thus better performance. Since the mid-1990s, k-ary n-cube networks were used by several high-performance multiprocessors such as the SGI Origin 2000 hypercube [43], the 2-D torus of the Cray X1 [16], the 3-D torus of the Cray T3E [55] and XT3 [12, 17] and the torus of the Alpha 21364 [49] and IBM BlueGene [35]. However, the increasing pin bandwidth has recently motivated the migration towards *high*-radix topologies such as the radix-64 folded-Clos topology used in the Cray BlackWidow system [56]. In this chapter, we will discuss mesh/torus topologies while in the next chapter, we will present high-radix topologies.

3.2   TYPES OF NETWORKS

Topologies can be broken down into two different genres: direct and indirect [20]. A *direct* network has processing nodes attached directly to the switching fabric; that is, the switching fabric is distributed among the processing nodes. An *indirect* network has the endpoint network independent of the endpoints themselves – i.e., dedicated switch nodes exist and packets are forwarded *indirectly* through these switch nodes. The type of network determines some of the packaging and cabling requirements as well as fault resilience. It also impacts cost, for example, since a direct network can combine the switching fabric and the network interface controller (NIC) functionality in the same silicon package. An indirect network typically has two separate chips, with one for the NIC and another for the switching fabric of the network. Examples of direct network include mesh, torus, and hypercubes discussed in this chapter as well as high-radix topologies such as the flattened butterfly described in the next chapter. Indirect networks include conventional butterfly topology and fat-tree topologies.

The term *radix* and *dimension* are often used to describe both types of networks but have been used differently for each network. For an indirect network, *radix* often refers to the number of ports of a switch, and the *dimension* is related to the number of stages in the network. However, for a direct network, the two terminologies are reversed – *radix* refers to the number of nodes within a dimension, and the network size can be further increased by adding multiple *dimensions*. The two terms are actually a duality of each other for the different networks – for example, in order to reduce the network diameter, the *radix* of an indirect network or the *dimension* of a direct network can be increased. To be consistent with existing literature, we will use the term *radix* to refer to different aspects of a direct and an indirect network.

3.3   MESH, TORUS, AND HYPERCUBES

The *mesh*, *torus* and *hypercube* networks all belong to the same family of direct networks often referred to as k-ary n-mesh or k-ary n-cube. The scalability of the network is largely determined by the radix, k, and number of dimensions, n, with $N = k^n$ total endpoints in the network. In practice, the radix of the network is not necessarily the same for every dimension (Figure 3.2). Therefore, a more general way to express the total number of endpoints is given by Equation 3.1.

$$N = \prod_{i=0}^{n-1} k_i \qquad (3.1)$$

(a) 8-ary 1-mesh.   (b) 8-ary 1-cube.

Figure 3.1: Mesh (a) and torus (b) networks.

Mesh and torus networks (Figure 3.1) provide a convenient starting point to discuss topology tradeoffs. Starting with the observation that each router in a k-ary n-mesh, as shown in Figure 3.1(a), requires only three ports; one port connects to its neighboring node to the left, another to its right neighbor, and one port (not shown) connects the router to the processor. Nodes that lie along the edge of a mesh, for example nodes 0 and 7 in Figure 3.1(a), require one less port. The same applies to k-ary n-cube (torus) networks. In general, the number of input and output ports, or *radix* of each router is given by Equation 3.2. The term "radix" is often used to describe *both* the number of input and output ports on the router, and the size or number of nodes in each dimension of the network.

$$r = 2n + 1 \tag{3.2}$$

The number of dimensions (n) in a mesh or torus network is limited by practical packaging constraints with typical values of n=2 or n=3. Since n is fixed we vary the radix (k) to increase the size of the network. For example, to scale the network in Figure 3.2a from 32 nodes to 64 nodes, we increase the radix of the y dimension from 4 to 8 as shown in Figure 3.2b.

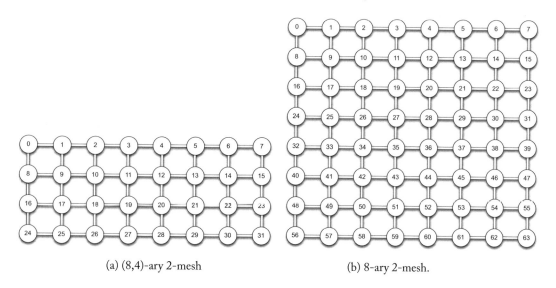

(a) (8,4)-ary 2-mesh

(b) 8-ary 2-mesh.

Figure 3.2: Irregular (a) and regular (b) mesh networks.

Since a binary hypercube (Figure 3.4) has a fixed radix (k=2), we scale the number of dimensions (n) to increase its size. The number of dimensions in a system of size N is simply $n = lg_2(N)$ from Equation 3.1.

$$r = n + 1 = lg_2(N) + 1 \tag{3.3}$$

As a result, hypercube networks require a router with more ports (Equation 3.3) than a mesh or torus. For example, a 512 node 3-D torus (n=3) requires seven router ports, but a hypercube requires $n = lg_2(512) + 1 = 10$ ports. It is useful to note, an n-dimension binary hypercube is isomorphic to a $\frac{n}{2}$-dimension torus with radix 4 ($k=4$). Router *pin bandwidth* is limited, thus building a 10-ported router for a hypercube instead of a 7-ported torus router may not be feasible without making each port *narrower*.

3.3.1   NODE IDENTIFIERS

The nodes in a k-ary n-cube are identified with an n-digit, radix k number. It is common to refer to a node identifier as an endpoint's "network address." A packet makes a finite number of hops in each of the n dimensions. A packet may traverse an intermediate router, c_i, en route to its destination. When it reaches the correct ordinate of the destination, that is $c_i = d_i$, we have resolved the ith dimension of the destination address.

3.3.2 k-ARY n-CUBE TRADEOFFS

The *worst-case* distance (measured in hops) that a packet must traverse between any source and any destination is called the *diameter* of the network. The network diameter is an important metric as it bounds the worst-case latency in the network. Since each hop entails an arbitration stage to choose the appropriate output port, reducing the network diameter will, in general, reduce the *variance* in observed packet latency. The network diameter is independent of traffic pattern, and is entirely a function of the topology, as shown in Table 3.1

Table 3.1: Network diameter and average latency.

Network	Diameter (hops)	Average (hops)
mesh	$k - 1$	$(k + 1)/3$
torus	$k/2$	$k/4$
hypercube	n	$n/2$
flattened butterfly	$n + 1$	$n + 1 - (n - 1)/k$

from/to	0	1	2	3	4	5	6	7	8
0	0	1	2	3	4	5	6	7	8
1	1	0	1	2	3	4	5	6	7
2	2	1	0	1	2	3	4	5	6
3	3	2	1	0	1	2	3	4	5
4	4	3	2	1	0	1	2	3	4
5	5	4	3	2	1	0	1	2	3
6	6	5	4	3	2	1	0	1	2
7	7	6	5	4	3	2	1	0	1
8	8	7	6	5	4	3	2	1	0

(a) radix-9 mesh

from/to	0	1	2	3	4	5	6	7	8
0	0	1	2	3	4	4	3	2	1
1	1	0	1	2	3	4	4	3	2
2	2	1	0	1	2	3	4	4	3
3	3	2	1	0	1	2	3	4	4
4	4	3	2	1	0	1	2	3	4
5	4	4	3	2	1	0	1	2	3
6	3	4	4	3	2	1	0	1	2
7	2	3	4	4	3	2	1	0	1
8	1	2	3	4	4	3	2	1	0

(b) radix-9 torus

Figure 3.3: Hops between every source, destination pair in a mesh (a) and torus (b).

In a mesh (Figure 3.3), the destination node is, at most, k-1 hops away. To compute the average, we compute the distance from all sources to all destinations, thus a packet from node 1 to node 2 is one hop, node 1 to node 3 is two hops, and so on. Summing the number of hops from each source to each destination and dividing by the total number of packets sent $k(k-1)$ to arrive at the average hops taken. A packet traversing a torus network will use the *wraparound* links to reduce the average hop count and network diameter. The worst-case distance in a torus with radix k is $k/2$, but the average distance is only half of that, $k/4$. In practice, when the radix k of a torus is odd, and there are two equidistant paths regardless of the direction (i.e., whether the wraparound link is used) then a routing convention is used to break ties so that half the traffic goes in each direction across the two paths.

A binary hypercube (Figure 3.4) has a fixed radix ($k=2$) and varies the number of dimensions (n) to scale the network size. Each node in the network can be viewed as a binary number, as shown in Figure 3.4. Nodes that differ in only one digit are connected together. More specifically, if two nodes differ in the ith digit, then they are connected in the ith dimension. Minimal routing in a hypercube will require, at most, n hops if the source and destination differ in every dimension, for example, traversing from 000 to 111 in Figure 3.4. On average, however, a packet will take $n/2$ hops.

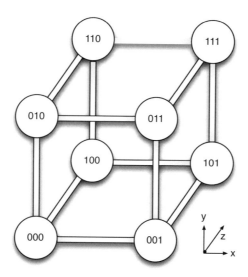

Figure 3.4: A binary hypercube with three dimensions.

SUMMARY

This chapter provided an overview of direct and indirect networks, focusing on topologies built from *low-radix* routers with a relatively small number of *wide* ports. We describe key performance metrics of *diameter* and *average hops* and discuss tradeoffs. Technology trends motivated the use of low-radix topologies in the 80s and the early 90s.

In practice, there are other issues that emerge as the system architecture is considered as a whole; such as, QoS requirements, flow control requirements, and tolerance for latency variance. However, these are secondary to the guiding technology (signaling speed) and packaging and cooling constraints. In the next chapter, we describe how evolving technology motivates the use of *high-radix* routers and how different high-radix topologies can efficiently exploit these *many-ported* switches.

CHAPTER 4

High-Radix Topologies

Dally [18] and Agarwal [5] showed that under fixed packaging constraints, lower radix networks offered lower packet latency. As a result, many studies have focused on low-radix topologies such as the k-ary n-cube topology discussed in Chapter 3. The fundamental result of these authors still holds – technology and packaging constraints should drive topology design. However, what has changed in recent years are the topologies that these constraints lead us toward. In this section, we describe the high-radix topologies that can better exploit today's technology.

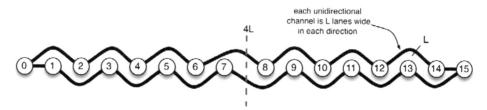

(a) radix-16 one-dimensional torus with each unidirectional link L lanes wide.

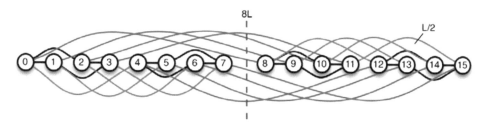

(b) radix-4 two-dimensional torus with each unidirectional link $L/2$ lanes wide.

Figure 4.1: Each router node has the same amount of pin bandwidth but differ in the number of ports.

4.1 TOWARDS HIGH-RADIX TOPOLOGIES

Technology trends and packaging constraints can and do have a major impact on the chosen topology. For example, consider the diagram of two 16-node networks in Figure 4.1. The radix-16 one-dimensional torus in Figure 4.1a has *two* ports on each router node; each port consists of an input and output and are *L lanes* wide. The amount of *pin bandwidth* off each router node is $4 \times L$. If we partitioned the router bandwidth slightly differently, we can make better use of the bandwidth as shown in Figure 4.1b. We transformed the one-dimensional torus of Figure 4.1a into a radix-4 two-dimensional torus in Figure 4.1b, where we have *twice* as many ports on each router, but each port is only *half* as wide — so the pin bandwidth on the router is held constant. There are several direct benefits of the *high-radix* topology in Figure 4.1b compared to its *low-radix* topology in Figure 4.1a:

(a) by increasing the number of *ports* on each router, but making each port *narrower*, we doubled the amount of *bisection bandwidth*, and

(b) we decreased the average number of hops by half.

The topology in Figure 4.1b requires longer cables which can adversely impact the signaling rate since the maximum bandwidth of an electrical cable drops with increasing cable length since signal attenuation due to skin effect and dielectric absorption increases linearly with distance.

4.2   TECHNOLOGY DRIVERS

The trend toward high-radix networks is being driven by several technologies:

- high-speed signaling, allowing each channel to be *narrower* while still providing the same bandwidth,

- affordable optical signaling through CMOS photonics and *active* optical cables that decouple data rate from cable reach, and

- new router microarchitectures that scale to high port counts and exploit the abundant wire and transistor density of modern CMOS devices.

The first two items are described further in this section while the router microarchitecture details will be discussed in Chapter 6.

4.2.1   PIN BANDWIDTH

As described earlier in Chapter 2, the amount of total pin bandwidth has increased at a rate of $100\times$ over each decade for the past 20-25 years. To understand how this increased pin bandwidth affects the optimal network radix, consider the latency (T) of a packet traveling through a network. Under low loads, this latency is the sum of header latency and serialization latency. The header latency (T_h) is the time for the beginning of a packet to traverse the network and is equal to the number of hops (H) a packet takes times a per hop router delay (t_r). Since packets are generally wider than the network channels, the body of the packet must be squeezed across the channel, incurring an additional serialization delay (T_s). Thus, total delay can be written as

$$T = T_h + T_s = Ht_r + L/b \tag{4.1}$$

where L is the length of a packet, and b is the bandwidth of the channels. For an N node network with radix k routers (k input channels and k output channels per router), the number of hops[1] must be at least $2 log_k N$. Also, if the total bandwidth of a router is B, that bandwidth is divided among the $2k$ input and output channels and $b = B/2k$. Substituting this into the expression for latency from Equation (4.1)

$$T = 2t_r \log_k N + 2kL/B \qquad (4.2)$$

Then, setting dT/dk equal to zero and isolating k gives the optimal radix in terms of the network parameters,

$$k \log^2 k = \frac{Bt_r \log N}{L} \qquad (4.3)$$

In this differentiation, we assume B and t_r are independent of the radix k. Since we are evaluating the optimal radix for a *given* bandwidth, we can assume B is independent of k. The t_r parameter is a function of k but has only a small impact on the total latency and has no impact on the optimal radix. Router delay t_r can be expressed as the number of pipeline stages (P) times the cycle time (t_{cy}). As radix increases, the router microarchitecture can be designed where t_{cy} remains constant and P increases logarithmically. The number of pipeline stages P can be further broken down into a component that is independent of the radix X and a component which is dependent on the radix $Y \log_2 k$.[2] Thus, router delay (t_r) can be rewritten as

$$t_r = t_{cy} P = t_{cy}(X + Y \log_2 k) \qquad (4.4)$$

If this relationship is substituted back into Equation (4.2) and differentiated, the dependency on radix k coming from the router delay disappears and does not change the optimal radix. Intuitively, although a single router delay increases with a $\log(k)$ dependence, the effect is offset in the network by the fact that the hop count decreases as $1/\log(k)$ and as a result, the router delay does not significantly affect the optimal radix.

In Equation (4.2), we also ignore time of flight for packets to traverse the wires that make up the network channels. The time of flight does not depend on the radix(k) and thus has minimal impact on the optimal radix. Time of flight is D/v where D is the total physical distance traveled by a packet, and v is the propagation velocity. As radix increases, the distance between two router nodes increases. However, the *total* distance traveled by a packet will be approximately equal since the lower-radix network requires more hops.[3]

From Equation (4.3), we refer to the quantity $A = \frac{Bt_r \log N}{L}$ as the *aspect ratio* of the router [42]. This aspect ratio impacts the router radix that minimizes network latency. A high aspect ratio implies a "tall, skinny" router (many, narrow channels) minimizes latency, while a low ratio implies a "short, fat" router (few, wide channels).

---

[1]Uniform traffic is assumed and $2 log_k N$ hops are required for a non-blocking network.
[2]For example, routing pipeline stage is often independent of the radix while the switch allocation is dependent on the radix.
[3]The time of flight is also dependent on the packaging of the system but we ignore packaging in this analysis.

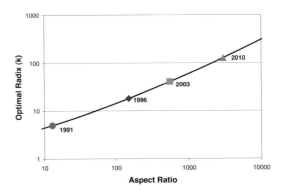

Figure 4.2: Relationship between the optimal radix for minimum latency and router aspect ratio. The labeled points show the approximate aspect ratio for a given year's technology with a packet size of L=128 bits

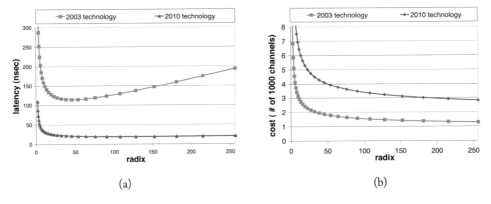

Figure 4.3: Latency (a) and cost (b) of the network as the radix is increased for two different technologies.

A plot of the minimum latency radix versus aspect ratio is shown in Figure 4.2 annotated with aspect ratios from several years. These particular numbers are representative of large supercomputers with single-word network accesses[4], but the general trend of the radix increasing significantly over time remains. Figure 4.3(a) shows how latency varies with radix for 2003 and 2010 aspect ratios. As radix is increased, latency first decreases as hop count, and hence T_h, is reduced. However, beyond a certain radix, serialization latency begins to dominate the overall latency and latency increases. As bandwidth, and hence aspect ratio, is increased, the radix that gives minimum latency also increases. For 2004 technology (aspect ratio = 652), the optimum radix is 45 while for 2010 technology (aspect ratio = 3013) the optimum radix is 128.

---

[4]The 1996 data is from the Cray T3E [55] (B=48Gb/s, t_r=40ns, N=2048), the 2003 data is combined from the Alpha 21364 [49] and Velio VC2002 [20] (1Tb/s, 10ns, 4096), and the 2010 data was estimated as (20Tb/s, 2ns, 8192).

Increasing the radix of networks monotonically reduces the overall cost of a network. Network cost is largely due to router pins and connectors and hence is roughly proportional to total router bandwidth: the number of channels times their bandwidth. For a fixed network bisection bandwidth, this cost is proportional to hop count. Since increasing radix reduces hop count, higher radix networks have lower cost as shown in Figure 4.3(b). Power dissipated by a network also decreases with increasing radix. The network power is roughly proportional to the number of router nodes in the network. As radix increases, hop count decreases, and the number of router nodes decreases. The power of an individual router node is largely independent of radix as long as total router bandwidth is held constant. Router power is largely due to SerDes (serializer/deserializer) I/O circuits and internal switch datapaths. The arbitration logic, which becomes more complex as radix increases, represents a negligible fraction of total power [67].

4.2.2   ECONOMICAL OPTICAL SIGNALING

Migrating from low-radix topology to high-radix topology increases the length of the channels as described earlier in Section 4.1. For low-radix routers, the routers are often only connected to neighboring routers – e.g., with a radix-6 router in a 3-D torus network, each router is connect to two neighbors in the x, y, and z dimensions. The long wraparound link of a torus topology can be removed by creating a "folded" torus, as shown in Figure 4.1(a). As a result, the cable lengths are reasonably short and only need to cross one or two cabinets at most and thus often under a few meters in length. The benefit of short cables, under say five meters, is that they can be driven using low-cost *passive* electrical signaling. With a high-radix router, such as a radix-64 router, each router is now connected to a larger number of routers which can be either centrally located or physically distributed, yet far away. Although high-radix reduces the network diameter, it increases the length of the cables required in the system as demonstrated in Figure 4.1(b).

Historically, the high cost of optical signaling limited its use to very long distances or applications that demanded performance regardless of cost. Recent advances in silicon photonics and their application to active optical cables such as Intel Connects Cables [23] and Luxtera Blazar [46, 47] have enabled economical optical interconnect. These active optical cables have electrical connections at either end and EO and OE [5] modules integrated into the cable itself.

Figure 4.4 compares the cost of electrical and optical signaling bandwidth as a function of distance. The cost of Intel Connects Cables[23] is compared with the electrical cable cost model presented in [41]. [6] Optical cables have a higher fixed cost (y-intercept) but a lower cost per unit distance (slope) than electrical cables. Based on the data presented here, the crossover point is at 10m. For distances shorter than 10m, electrical signaling is less expensive. Beyond 10m, optical signaling is more economical. By reducing the number of global cables it minimizes the effect of the higher fixed overhead of optical signaling, and by making the global cables longer, it maximizes

---

[5] EO : Electrical to Optical, OE : Optical to Electrical
[6] The optical cost was based on prices available at http://shop.intel.com at the time this analysis was done in 2008 [38]. If purchased in bulk, the prices will likely be lower. The use of multi-mode fiber instead of single-mode fiber may also result in lower cost. Subsequently, the Connects Cables were acquired from Intel by EMCORE Corporation.

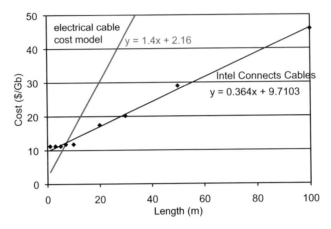

Figure 4.4: Cable cost comparison between optical and electrical cables.

the advantage of the lower per-unit cost of optical fibers. The high-radix topologies described in the following section exploits this relationship between cost and distance and thus, exploiting the availability of high-radix routers.

4.3   HIGH-RADIX TOPOLOGY

4.3.1   HIGH-DIMENSION HYPERCUBE, MESH, TORUS

The direct networks described earlier in Chapter 3 can use high-radix routers to create high-*dimension* topologies, including hypercube, mesh, and torus. The high-dimension topologies reduce the network diameter, but since the number of routers required for these topologies is proportional to N or the network size, the wiring or the cabling complexity can become prohibitively expensive and also increase the network cost. The indirect networks described earlier in Chapter 3 can better exploit high-radix routers while reduce network cost and wiring complexity. In addition, concentration [20] can be used to where the router is shared among multiple terminal nodes to further reduce the wiring complexity. The topologies that we describe in this chapter leverage concentration to exploit high-radix routers (which enable connecting multiple nodes to a router) and make cabling feasible.

4.3.2   BUTTERFLY

The butterfly network (k-ary n-fly) can take advantage of high-radix routers to reduce latency and network cost [20]. For a network with N nodes and radix-k routers, $log_k(N) + 1$ stages with N/k routers in each stage are needed. For example, a 64-node butterfly network with radix-4 routers (4-ary 3-fly) is shown in Figure 4.5, with the input nodes shown on the left and the output nodes shown on the right. The butterfly topology minimizes the network diameter and as a result, minimizes network cost. However, there are two noticeable disadvantages of the butterfly network. There is a lack of

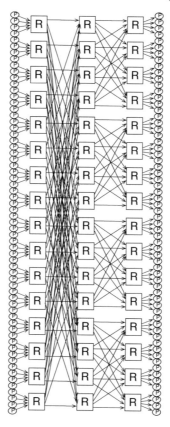

Figure 4.5: Conventional Butterfly Topology (4-ary 3-fly) with 64 nodes. *P* represents the processor or the terminals nodes and *R* represents the switches or the routers. For simplicity, the network injection ports (terminal ports) are shown on the left while the network ejection ports are shown on the right. However, they represent the same physical terminal nodes.

path diversity in the topology as there is only a single path between any source and any destination. This results in poor throughput for any non-uniform traffic patterns. In addition, a butterfly network cannot exploit traffic locality as all packets must traverse the diameter of the network.

4.3.3   HIGH-RADIX FOLDED-CLOS

A Clos network [14] is a multi-stage interconnection network consisting of an odd number of stages connecting input ports to output ports (Figure 4.6(a)). The Clos network can be created by combining two butterfly networks back-to-back with the first stage used for load-balancing (input network) and the second stage used to route the traffic (output network). A Clos network provides

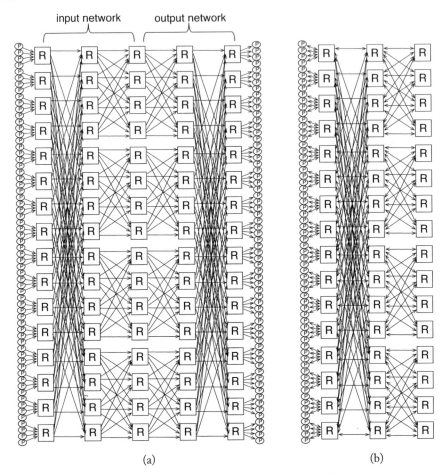

Figure 4.6: (a) High-Radix Clos topology and (b) corresponding folded-Clos topology. The channels in the folded-Clos represent bidirectional channels. The routers in the right-most column of (b) are radix-4 while the others in (b) are radix-8. If the same radix-8 routers need to be used in the folded-Clos network, two radix-4 routers can be combined into a single radix-8 router.

many paths – one for each middle-stage switch in the Clos – between each pair of nodes. This path diversity enables the Clos to route arbitrary traffic patterns with no loss of throughput.

Like a butterfly (k-ary n-fly) network, a folded-Clos is also an *indirect* network. The router nodes are distinct from the endpoints. The first tier of the network connects $k/2$ hosts (endpoints) to the switch, and $k/2$ *uplinks* to other switches in the next tier. If the injection bandwidth is balanced with the uplink bandwidth, we refer to the network as *fully provisioned*; however, if there is more injection bandwidth than uplink bandwidth, then it is *oversubscribed*. Oversubscription is common in datacenter applications since it reduces cost and improves utilization. The input and output stages

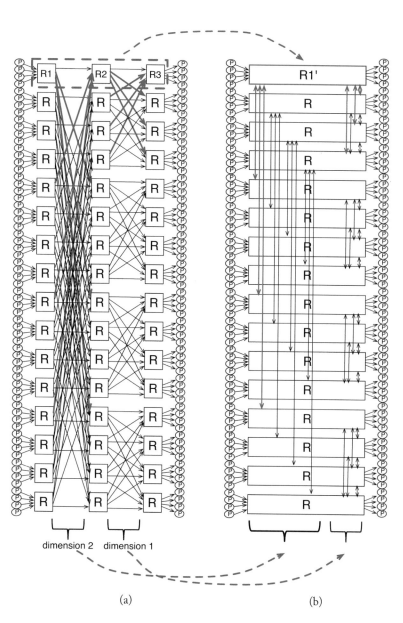

(a)   (b)

Figure 4.7: Deriving a Flattened Butterfly from a conventional butterfly shown in Figure 4.5.

of a Clos network can be combined or *folded* on top of one another creating a folded Clos or fat-tree [44] network which can exploit traffic locality with the input/output ports co-located, as shown in Figure 4.6(b).

A Clos or folded Clos network, however, has a cost that is nearly double that of a butterfly with equal capacity and has greater latency than a butterfly. The increased cost and latency both stem from the need to route packets first to an arbitrary middle stage switch and then to their ultimate destination. This doubles the number of long cables in the network, which approximately doubles cost, and doubles the number of inter-router channels traversed, which drives up latency.

4.3.4 FLATTENED BUTTERFLY

To overcome the limitations of the folded-Clos topology, the flattened butterfly [41] removes intermediate stages and creates a *direct* network. As a result, the flattened butterfly is a topology that exploits high-radix routers to realize lower cost than a Clos on load-balanced traffic, and provide better performance and path diversity than a conventional butterfly. The flattened butterfly can be derived by starting with a conventional butterfly (*k*-ary *n*-fly) and combining or *flattening* the routers in each *row* of the network into a single router. An example of flattened butterfly construction is shown in Figure 4.7. 4-ary 3-fly network is shown in Figure 4.7(a) with the corresponding flattened butterflies shown in Figure 4.7(b). The routers R1, R2, and R3 from the first row of Figure 4.7(a) are combined into a single router R0' in the flattened butterfly of Figure 4.7(b). As a row of routers is combined, channels entirely local to the row, e.g., channel (R0,R1) in Figure 4.7(a), are eliminated. All other channels of the original butterfly remain in the flattened butterfly. Because channels in a flattened butterfly are symmetrical, each line in Figures 4.7(b) represents a bidirectional channel (i.e., two unidirectional channels), while each line in Figures 4.7(a) represents a unidirectional channel.

A *k*-ary *n*-flat, the flattened butterfly derived from a *k*-ary *n*-fly, is composed of $\frac{N}{k}$ radix $k' = n(k - 1) + 1$ routers where N is the size of the network. The routers are connected by channels in $n' = n - 1$ dimensions, corresponding to the $n - 1$ columns of inter-rank wiring in the butterfly. In each dimension d, from 1 to n', router i is connected to each router j given by

$$j = i + \left[m - \left(\left\lfloor \frac{i}{k^{d-1}} \right\rfloor \bmod k \right) \right] k^{d-1} \tag{4.5}$$

for m from 0 to $k - 1$, where the connection from i to itself is omitted. For example, in Figure 4.7(d), R4' is connected to R5' in dimension 1, R6' in dimension 2, and R0' in dimension 3. With this construction, it is easy to see that the flattened butterfly is equivalent to the *generalized hypercube* topology [10], but with *k*-way concentration. With this concentration, the topology is better able to exploit the properties of high-radix routers.

4.3.5 DRAGONFLY

Although the flattened butterfly can cost-efficiently exploit high-radix routers, it is ultimately limited by the physical constraints of a router radix and cost of scaling to large node count. For example, if the router radix is limited to radix-64, the network can scale up to 64k nodes with three dimensions. However, to scale the network further, the number of dimensions of the flattened butterfly needs to be increased – which can create packaging difficulties as well as increase cost and latency. In addition, most of the channels (two of the three dimensions) require *global* or expensive channels which significantly increase the cost. To overcome this limitation, a collection of routers can be used together to create a very high-radix *virtual* router. The dragonfly topology [38] described in this section leverages this concept of a virtual router to create a more scalable topology.

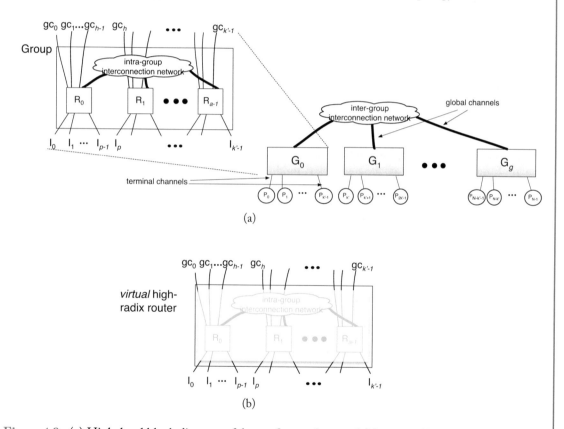

Figure 4.8: (a) High-level block diagram of dragonfly topology and (b) a virtual high-radix router.

The dragonfly is a hierarchical network with three levels: router, group, and system as shown in Figure 4.8. At the bottom level, each router has three different type of connections : 1) connections to p terminals, 2) $a - 1$ local channels to other routers in the same group, and 3) h global channels to routers in other groups. Hence, the radix (or degree) of each router is $k = p + a + h - 1$. A group consists of a routers connected via an intra-group interconnection network formed from local channels. Each group has ap connections to terminals and ah connections to global channels, and all of the routers in a group collectively act as a *virtual router* with radix $k' = a(p + h)$. As shown in

Figure 4.8(b), if the details of the intra-group is ignored, a group can be viewed as a virtual high-radix router. This very high radix, $k' >> k$ enables the system level network to be realized with very low global diameter (the maximum number of expensive global channels on the minimum path between any two nodes). Up to $g = ah + 1$ groups ($N = ap(ah + 1)$ terminals) can be connected with a global diameter of one. In contrast, a system-level network built directly with radix k routers would require a larger global diameter.

In a maximum-size ($N = ap(ah + 1)$) dragonfly, there is exactly one connection between each pair of groups. In smaller dragonflies, there are more global connections out of each group than there are other groups. These excess global connections are distributed over the groups with each pair of groups connected by at least $\lfloor \frac{ah+1}{g} \rfloor$ channels. The dragonfly parameters a, p, and h can have any values. However, to balance channel load on load-balanced traffic, the network should have $a = 2p = 2h$. Because each packet traverses two local channels along its route (one at each end of the global channel) for one global channel and one terminal channel, this ratio maintains balance. Additional details of routing and load-balancing on the dragonfly topology will be discussed in Chapter 5. Because global channels are expensive, deviations from this 2:1 ratio should be done in a manner that overprovisions local and terminal channels, so that the expensive global channels remain fully utilized. That is, the network should be balanced so that $a \geq 2h, 2p \geq 2h$.

The scalability of a balanced dragonfly is shown in Figure 4.9. By increasing the effective

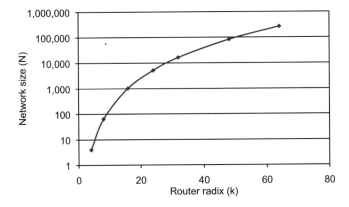

Figure 4.9: Scalability of the dragonfly topology as router radix increases. 1D flattened butterfly is assumed for both the intra- and the inter-group networks.

radix, the dragonfly topology is highly scalable – with radix-64 routers, the topology scales to over 256k nodes with a network diameter of only three hops. In comparison, a 2D flattened butterfly using radix-64 routers can scale to approximately 10k nodes while a 3D flattened butterfly can only scale up to 64k nodes. Arbitrary networks can be used for the intra-group and inter-group networks in Figure 4.8. However, to minimize the network cost, a flattened butterfly with the smallest number of dimensions will be appropriate. A simple example of the dragonfly is shown in Figure 4.10 with

$p = h = 2, a = 4$ that scales to $N = 72$ with $k = 7$ routers. For both the intra- and inter-group networks, a 1D flattened butterfly (or a fully connected topology) is used. By using virtual routers, the effective radix is increased from $k = 7$ to $k' = 16$.

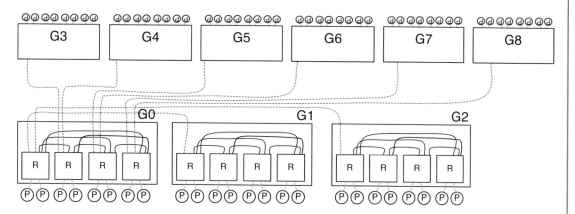

Figure 4.10: Example of a Dragonfly Topology for $N = 72$. The dotted line represents the intra-group channels and for simplicity, only the intra-group channels for $G0$ is shown.

4.3.6   HYPERX

The flattened butterfly described earlier in this chapter is *regular* as the number of nodes or switches in each dimension are identical. The HyperX [6] topology extends the flattened butterfly topology to create a more general class of topology. Similar to the flattened butterfly, all routers in each dimension of the HyperX are fully connected to other peers in each dimension. However, HyperX allows different number of switches in each dimension. An example of a HyperX topology is shown in Figure 4.11 with two dimensions ($L = 2$), different number of switches in each dimension ($S_1 = 2, S_2 = 4$), and a concentration of 4 ($T = 4$) as 4 terminals or end nodes are connected to each switch. Another parameter used to define a HyperX topology is K, which represents the relative bandwidth of the channels in each dimension, where the unit of bandwidth is the terminal bandwidth (i.e., the bandwidth between the terminals or the end nodes and the switch). The example in Figure 4.11 assumed $K = 1$ as all the channels had equal bandwidth. However, if $K = 2$, the inter-switch channels would have $2\times$ the bandwidth of the terminal channels. Different HyperX topologies can be described by these four parameters (L, S, K, T) and each value of S and K in each dimension can have the same value (i.e., $S_1 = S_2 = \ldots = s$) to create a *regular* HyperX or have different values for each dimensions and result in an *irregular* HyperX. Using these parameters, an n-dimensional hypercube can be described as ($L = n, S = 2, K = 1, T = 1$) and a k-ary n-flat flattened butterfly can be described as ($L = n, T = S = k, K = 2$).

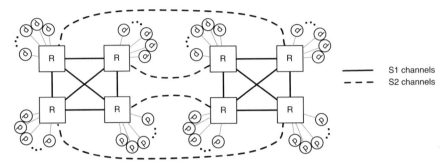

Figure 4.11: HyperX topology with two dimensions and different number of switches in each dimension. Dotted lines represent channels for switches in the same dimension 1 ($S1$) while solid lines represent channels in the same dimension 2 ($S2$).

SUMMARY

This chapter provides an introduction into how the evolving technology motivates the migration towards *high*-radix networks, compared to previous *low*-radix networks. Different high-radix topologies that have been recently proposed are also presented which include flattened butterfly, dragonfly, and HyperX. The optimal topology for a given large-scale system is ultimately determined by the packaging constraint and packaging/signaling cost. For example, dragonfly [38] was shown to be more cost-efficient based on the cost model described in Figure 4.4; however, if the cost of active cables are significantly reduced further such that they are similar to the cost of electrical cables, a more *flattened* topology such as the flattened butterfly or the HyperX, instead of a hierarchical topology such as the dragonfly, can result in a more cost-efficient topology. To fully exploit the benefits of these topologies compared to other high-radix topologies such as the folded-Clos topology, proper adaptive routing is critical to achieve the full benefits of these topology. In the next chapter, we describe the different types of routing algorithms that can be implemented on these topologies to take advantage of the path diversity, both minimal and nonminimal paths to load-balance the channels.

CHAPTER 5

Routing

Routing determines the path a packet takes from its source to its destination. Even if the topology provides path diversity, it is the routing algorithm that determines whether the path diversity is exploited or not. In addition, proper routing algorithms are critical to fully exploit the benefits of the recently proposed topologies such as the flattened butterfly or the dragonfly topology described in the previous chapter. In this section, we review the basics of routing algorithm in interconnection networks and present routing algorithms on recently proposed high-radix topologies.

5.1 ROUTING BASICS

Routing algorithms can be classified according to the following different metrics:

- Adaptivity:

  - **Adaptive Routing:** The state of the network is incorporated in making the routing decision to adapt to network state such as network congestion.

  - **Oblivious Routing:** No network information is used in making the routing decision. Deterministic routing can also be classified as oblivious routing.

- Hop Count:

  - **Minimal Routing:** Minimal number of hop count between source and destination is traversed. Depending on the topology and the adaptivity of the routing algorithm, there might be multiple minimal paths.

  - **Nonminimal Routing:** The number of hop count traversed enroute to the destination node exceeds the minimal hop count. Nonminimal routing increases path diversity and can improve network throughput.

- Routing Decision:

  - **Source routing:** The routing path is determined at the source and the *path computation* only needs to be done once for each packet.

  - **Per-hop routing:** At each hop enroute to the destination, the packet goes through routing computation to determine the next productive hop. The progressive adaptive routing (PAR) described in this chapter is a variant of per-hop routing, and the adaptive routing decision is revisited at each hop.

- Routing Implementation: Regardless of the routing decision mechanism, a *route computation* is needed to determine the output port at each router. The route computation can be implemented using either algorithmic logic or a table-based structure.

  - **Algorithmic:** Based on the current node and destination information, a fixed logic can be used to determine the output port. This can result in simple logic but inflexible routing algorithm.
  - **Table-based routing:** A lookup table can be implemented whose inputs are either the source (or current node) and destination, and the table returns the appropriate output port.

5.1.1   OBJECTIVES OF A ROUTING ALGORITHM

In designing a routing algorithm for a given topology, the objective of the algorithm should include the following:

- **path diversity** : Exploit the path diversity of the topology, which can include both minimal and non-minimal paths.

- **load balancing** : Proper load-balancing of the channels across both benign and adversarial traffic pattern is needed to achieve high throughput.

- **complexity effective** : To minimize the impact of the routing algorithm on packet latency, and load imbalance that may result from a fault in the network, the routing algorithm must be able to be implemented efficiently. For example, it is not practical for a routing algorithm to perform a simulated-annealing process to find the optimal load balance each time a network fault occurs.

In the rest of this chapter, we describe various routing algorithms that try to achieve these characteristics on both conventional topologies (Chapter 3) as well as recently proposed high-radix topologies (Chapter 4).

5.2   MINIMAL ROUTING

With minimal routing, all packets traverse the minimal hop count from source to its destination. Minimal routing can be done either deterministically, obliviously, or adaptively.

5.2.1   DETERMINISTIC ROUTING

The simplest form of minimal routing is using dimension-ordered routing (DOR) where the routing is restricted to traverse in a pre-determined order. For example, XY routing can be used on a 2D mesh network where all packets first traverse in the X dimension and then, traverse the Y dimension to reach its destination. This is the simplest form of routing in a given topology but does not exploit any possible path diversity and can not load-balance channels on adversarial traffic patterns.

5.2.2   OBLIVIOUS ROUTING

If path diversity exists in the topology (i.e., there is more than one minimal path between a source and a destination), oblivious minimal routing can be used and take advantage of path diversity. All routing paths consist of minimal hop count but different paths can be used according to the routing algorithm. Examples of minimal oblivious routing include ROMM [50], O1turn [57], and CDR [3].

ROMM (Randomized, Oblivious Multi-phase Minimal) routing consists of p phases and $p - 1$ randomly selected intermediate nodes. The packets are routed to an intermediate node in each of the first $p - 1$ phases and routed to its destination in the final phase. The intermediate nodes are selected such that the routing is still minimal – i.e., within each phase, the routing output at each route is *productive* as the packet moves closer to the destination. To avoid routing deadlock, p virtual channels (VC) [21] are needed. While ROMM provides very high path diversity, randomized DOR or O1turn routing [57] on a 2D mesh network limits path diversity to 2. For each packet, O1turn routing algorithm randomly select either XY or YX routing – with a probability of 1/2, packets are routed using XY DOR while the other packets are routed using YX DOR. This routing algorithm maintains the simplicity of a DOR algorithm except for the need for an extra VC. Packets routed using XY use one VC while packets routed using YX use another VC to avoid routing deadlock. Despite providing only a path diversity of 2, it has been shown to be near-optimal in its performance [57].

A variation of O1turn routing algorithm is the class-based deterministic routing (CDR) [3] algorithm. Similar to O1turn, both XY and YX routing is used but instead of randomly selecting a packet for either XY or YX routing, the routing path is determined by the packet class. For example, if there are request and reply traffic class in the network, packets in the request class use one routing algorithm (i.e., XY) while the other class (reply traffic) uses YX routing. CDR exploits path diversity of the topology while all packets still traverse minimal hop count. Compared to O1turn, this routing algorithm has the additional benefit of reducing the number of VCs needed since another separate set of VC are not need to avoid protocol deadlock as the same VCs can be used for both routing and protocol deadlock avoidance.

5.3   NON-MINIMAL ROUTING

Minimal routing minimizes the hop count but when network congestion occurs, taking a non-minimal route can sometimes reduce the network latency. In addition, for adversarial traffic patterns, better load-balancing of the channels can be achieved with non-minimal routing and result in higher network throughput. In this section, we describe an oblivious non-minimal routing algorithm (Valiant's algorithm [66]) and several different adaptive non-minimal routing algorithm including Universal Globally Adaptive routing algorithm (UGAL) [58].

5.3.1   VALIANT'S ALGORITHM (VAL)

VAL routing [66] uses randomization and non-minimal routing to exploit the path diversity of a topology and achieve load-balancing. VAL is a two-phase routing algorithm with a random node in the network initially selected. In the first phase, minimal routing is used to route the packet to the intermediate node. Once the packet reaches the intermediate node, in the second phase, the packet is routed to the destination. If the random node happens to be either the source or the destination, VAL degenerates into minimal routing as only one phase of the VAL routing is needed.

By using randomization to load-balance the traffic, high throughput can be achieved on adversarial traffic patterns as VAL can provide optimal performance on adversarial traffic pattern, or 50% throughput of capacity [63]. However, by converting all traffic pattern into two phases of uniform random traffic, VAL causes higher zero-load latency and loss of traffic locality. Thus, on benign traffic pattern such as uniform random traffic, network throughput is reduced by a factor of 2 compared to minimal routing. In the following sections, we discuss how adaptive routing can be used to adaptively decide between minimal and nonminimal routing to maximize performance to overcome the limitations of VAL routing.

5.3.2   UNIVERSAL GLOBAL ADAPTIVE LOAD-BALANCING (UGAL)

UGAL routing [58] was proposed to adapt between minimal and nonminimal routing on a per-packet basis based on the congestion information of the network. If nonminimal routing is chosen, packet is routed as VAL routing – otherwise, minimal routing is used. For benign traffic patterns, UGAL attempts to approach the performance of minimal routing to exploit traffic locality or benign traffic patterns. For adversarial traffic patterns, UGAL sends most of its traffic nonminimally using VAL to load-balance the channels.

The UGAL routing decision is based at the *source* router – i.e., the router connected to the source node of the packet – and once the routing decision is made, the routing decision is not revisited as the packet follows either the minimal or the nonminimal path. The congestion information used by the UGAL routing algorithm is the product of the queue depth (q) and the hop count (H) for a minimal and a nonminimal path. The minimal queue depth (q_m) represents the congestion on the output port which is used for minimal routing and minimal hop count (H_m) is the minimal hop count between source and destination. With a randomly chosen intermediate node, nonminimal queue depth (q_{nm}) represents the congestion on the output port which is used for minimal routing to reach the randomly selected intermediate node while nonminimal hop count (H_{nm}) is the sum of the hop count from the source to the intermediate node and the hop count from the intermediate node to the destination. Thus, UGAL can be summarized as follows:

```
if   (q_m H_m ≤ q_nm H_nm)

        route minimally;

else

        route nonminimally;
```

Other routing algorithms such as Globally Oblivious Adaptive Local (GOAL) [59] routing or Channel-Queue Routing (CQR) [60] implement similar source-based adaptive routing algorithm in a torus network such that minimal or nonminimal routing decision is made at the source router.

5.3.3   PROGRESSIVE ADAPTIVE ROUTING (PAR)

Unlike UGAL where adaptive routing decision is made only once at the source router, *incremental* adaptive routing can be done using Progressive Adaptive Routing (PAR) [34] where the adaptive routing decision is revisited at each hop. By incrementally adapting, a better sense of congestion can be obtained – i.e., a congestion might not be observed at the source router but as the packet traverses the network, congestion can be encountered. PAR attempts to avoid this limitation of source adaptive routing by progressively re-evaluating the adaptive routing decision. However, to ensure forward progress and prevent livelock, restrictions in the amount of adaptivity are needed. In PAR routing, once a packet decides to route nonminimally (i.e., congestion is encountered), the packet is routed non-minimally without revisiting the adaptive routing decision.

5.3.4   DIMENSIONALLY-ADAPTIVE, LOAD-BALANCED (DAL) ROUTING

Similar to PAR, DAL [6] routing also attempts to adapt to congestion that is not visible at the source router and adapt *en route* to the destination. At each hop, all minimal and nonminimal paths are compared using only local congestion information. By revisiting the routing decision at each hop, a more accurate view of network congestion can be obtained and packets in-flight can switch from minimal to nonminimal as well as from nonminimal to minimal routing. To avoid livelock, restriction is applied such that packets can only be misrouted once per dimension. In addition, once a packet becomes *aligned* [1] with the destination in a given dimension, no misrouting is allowed for that particular dimension. Thus, DAL is able to *incrementally* select its nonminimal intermediate node based on congestion – instead of randomly selecting an intermediate node in the network as done with VAL implementation in adaptive routing algorithms such as UGAL.

5.4   INDIRECT ADAPTIVE ROUTING

Another class of adaptive routing that has been recently proposed is *indirect* adaptive routing [34, 39]. When congestion information used to make adaptive routing is not *directly* available at the source router, it becomes difficult to accurately estimate the congestion of the network. Congestion is often measured using queue lengths which relies on backpressure. However, if congestion information needs to propagate through multiple intermediate routers, backpressure needs to propagate through the intermediate routers which increases the propagation delay of the congestion information. Consider the dragonfly topology shown in Figure 5.1. Assume a packet in R1 is making its global adaptive routing decision of routing either minimally through gc_0 or non-minimally through gc_7. The rout-

---

[1]For a packet originating from source $(x_s, y_s, ..)$ to destination $(x_d, y_d, ..)$, for dimension i, if $i_s == i_d$, the packet is defined to be aligned with the destination in dimension i.

ing decision needs to load balance global channel utilization and ideally, the channel utilization can be obtained from the queues associated with the global channels, q_0 and q_3. However, q_0 and q_3 queue information are only available at R0 and R2 and not readily available at R1 – thus, the routing decision can only be made indirectly through the local queue information available at R1. In this example, q_1 reflects the state of q_0 and q_2 reflects the state of q_3. When either q_0 or q_3 is full, the flow control provides backpressure to q_1 and q_2 as shown with the arrows in Figure 5.1. As a result, in steady-state measurement, this local queue information can be used to accurately measure the throughput. Since the throughput is defined as the offered load when the latency goes to infinity (or the queue occupancy goes to infinity), this local queue information is sufficient. However, q_0 needs to be completely full in order for q_1 to reflect the congestion of gc_0 and allow R1 to route packets non-minimally. Thus, using local information requires sacrificing some packets to properly determine the congestion – resulting in packets being sent minimally having much higher latency. As the load increases, although minimally routed packets continue to increase in latency, more packets are sent non-minimally resulting in a decrease in average latency until saturation.

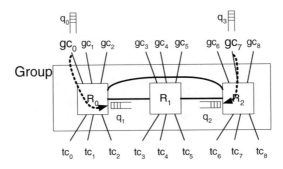

Figure 5.1: Example of routing *indirectness* in a dragonfly topology.

One implementation of indirect adaptive routing is the progressive adaptive routing described earlier in Section 5.3.3. Other implementations of indirect adaptive routing include using credit round-trip latency to stiffen backpressure [38], piggybacking congestion information [34], and using reservation-based routing mechanism [34]. As network size increases, *indirectness* becomes a more significant issue and properly incorporating it into the routing decision is critical to achieve the full benefit of adaptive routing.

5.5   ROUTING ALGORITHM EXAMPLES

In this section, we describe routing algorithms on different high-radix topologies, including the folded-Clos, flattened butterfly, and the dragonfly topologies. Even if the same routing algorithm is used, the different characteristics of a topology results in different implementations and different benefits of a routing algorithm.

5.5.1   EXAMPLE 1: FOLDED-CLOS

Routing a packet through a folded-Clos network proceeds in two phases: input and output. The input phase routing is also referred to as *uprouting* and the output phase routing is referred to as *downrouting*. During the input phase, a middle-stage switch is selected and the packet is routed to that switch. For a folded-Clos topology, the packet need not route all the way to the middle stage but can stop as soon as a common ancestor of the source and destination nodes is reached. Any middle-stage switch (or common ancestor switch) can be selected during the input phase. The selection may be made using either oblivious or adaptive routing. During the output phase, the packet is routed from the selected middle-stage switch (or common ancestor) to its destination output port. This routing is deterministic as there exists only a single path to the destination.

Many other topologies, including mesh, torus, or flattened butterfly, have path diversity that includes non-minimal paths. The folded-Clos topology also has high path diversity, but all of the paths are minimal. As a result, the difference between oblivious and adaptive routing is different from other topologies as both routing algorithms can exploit all the path diversity. For example, because the folded-Clos topology itself provides load-balancing capability with the *input* network as shown earlier in Figure 4.6(a), adaptive routing provides minimal benefit in terms of overall throughput on adversarial traffic patterns [40]. However, adaptive routing on a folded-Clos can provide benefits including the ability to route around nonuniformities in the network (such as the presence of deterministically routed traffic or faults) and provide lower variance in packet latency [40].

Recent router chips developed for high performance computing (HPC) systems that are often used in a fat-tree or a folded-Clos topology have included adaptive routing. For example, QsNet^{III} [53] implements adaptive routing in a fat-tree topology to adaptively select the common ancestor while also routing around faults in the network. Different adaptive routing strategies have also been proposed for Myrinet [24] to avoid the limitation of *static* or deterministic routing.

5.5.2   EXAMPLE 2: FLATTENED BUTTERFLY

Both minimal and nonminimal routing can be used on the flattened butterfly topology. If minimal routing is used along with deterministic routing, the flattened butterfly topology behaves identical to a conventional butterfly topology. An example of routing on a conventional butterfly is shown in Figure 5.2(a) with the corresponding minimal routing on the flattened butterfly in Figure 5.2(b) where the channels in the flattened butterfly are traversed in the same order as the conventional butterfly. However, since there are multiple dimensions, another possibly minimal path exists on the flattend butterfly as shown in Figure 5.3(a). If nonminimal routing is exploited, path diversity similar to a folded-Clos topology can be achieved. An example is shown in Figure 5.3(b) where the packet is first routed to an intermediate router (R13) before routing to its destination. In this section, we describe different routing algorithms on the flattened butterfly topology.

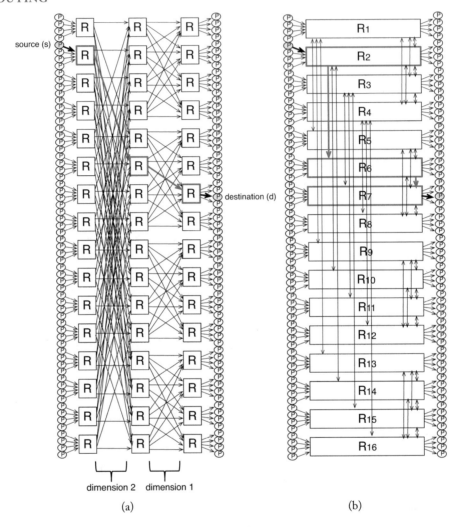

Figure 5.2: Routing example on (a) conventional butterfly and (b) corresponding flattened butterfly using minimal routing.

5.5.2.1 Minimal routing

Routing in a flattened butterfly requires a hop from a node to its local router, zero or more inter-router hops, and a final hop from a router to the destination node. If we label each node with a n-digit radix-k node address, an inter-router hop in dimension d changes the dth digit of the current node address to an arbitrary value, and the final hop sets the 0th (rightmost) digit of the current node address to an arbitrary value. Thus, to route minimally from node $a = a_{n-1}, \ldots, a_0$ to node $b = b_{n-1}, \ldots, b_0$ where a and b are n-digit radix-k node addresses involves taking one inter-router

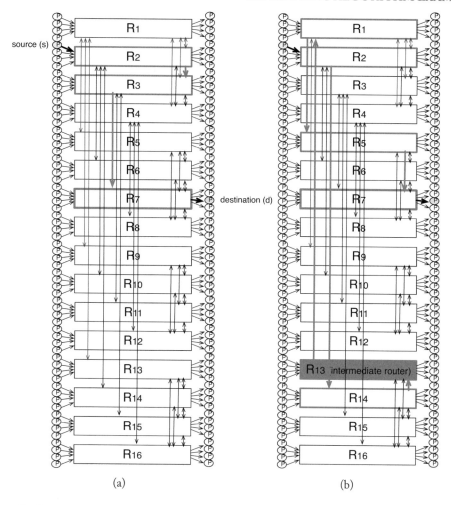

Figure 5.3: Different path diversity for the example shown earlier in Figure 5.2 with (a) minimal routing and (b) nonminimal routing.

hop for each digit, other than the rightmost, in which a and b differ. For example, in Figure 4.7(d) routing from node 0 (0000_2) to node 10 (1010_2) requires taking inter-router hops in dimensions 1 and 3. These inter-router hops can be taken in either order giving two minimal routes between these two nodes. In general, if two nodes a and b have addresses that differ in j digits (other than the rightmost digit), then there are $j!$ minimal routes between a and b. This path diversity derives from the fact that a packet routing in a flattened butterfly is able to traverse the dimensions in any order, while a packet traversing a conventional butterfly must visit the dimensions in a fixed order – leading to no path diversity.

Minimal Deterministic : The minimal deterministic algorithm chooses the next hop in a deterministic order – similar to dimension-ordered routing (DOR). An example of DOR on flattened butterfly includes routing packets along increasing dimension order. This routing algorithm does not require additional virtual channels (VCs) [21] for routing deadlock but no path diversity exists and results in a performance similar to a conventional butterfly.

Minimal Adaptive : The minimal adaptive algorithm operates by choosing for the next hop the productive channel with the shortest queue. To prevent deadlock, n' VC are used with the VC channel selected based on the number of hops remaining to the destination.

5.5.2.2 Non-minimal Routing

Routing non-minimally in a flattened butterfly provides additional path diversity and can achieve load-balanced routing for arbitrary traffic patterns. Consider, for example, Figure 4.7(b) and suppose that all of the traffic from nodes 0-3 (attached to router R0′) was destined for nodes 4-7 (attached to R1′). With minimal routing, all of this traffic would overload channel (R0′,R1′). By misrouting a fraction of this traffic to R2′ and R3′, which then forward the traffic on to R1′, load is balanced. With non-minimal routing, a flattened butterfly is able to match the load-balancing (and non-blocking) properties of a Clos network – in effect acting as a *flattened Clos*.

We consider routing in a k-ary n-flat where the source node s, destination node d, and current node c are represented as n-digit radix-k numbers, e.g., $s_{n-1}, \ldots, s_0$. At a given step of the route, a channel is productive if it is part of a minimal route; that is, a channel in dimension j is productive if $c_j \neq d_j$ before traversing the channel, and $c_j = d_j$ after traversing the channel.

Valiant (VAL) [66]: Valiant's algorithm load balances traffic by converting any traffic pattern into two phases of random traffic. It operates by picking a random intermediate node b, routing minimally from s to b, and then routing minimally from b to d. Routing through b perfectly balances load (on average) but at the cost of doubling the worst-case hop count, from n' to $2n'$. While any minimal algorithm can be used for each phase, our evaluation uses dimension order routing. Two VCs, one for each phase, are needed to avoid deadlock with this algorithm.

Universal Globally-Adaptive Load-balanced (UGAL [58], UGAL-S) : UGAL chooses between MIN AD and VAL on a packet-by-packet basis to minimize the estimated delay for each packet as described earlier in Section 5.3.2. With UGAL, traffic is routed minimally on benign traffic patterns and at low loads, matching the performance of MIN AD, and non-minimally on adversarial patterns at high loads, matching the performance of VAL. UGAL-S is identical to UGAL but with a *sequential* allocator while UGAL uses *greedy* allocator [40].

Adaptive Clos (CLOS AD): Like UGAL, the router chooses between minimal and non-minimal on a packet-by-packet basis using queue lengths to estimate delays. If the router chooses to route a packet non-minimally, however, the packet is routed as if it were adaptively routing to the middle stage of a Clos network. A non-minimal packet arrives at the intermediate node b by traversing each dimension using the channel with the shortest queue for that dimension (including a "dummy queue" for staying at the current coordinate in that dimension). Like UGAL-S, CLOS

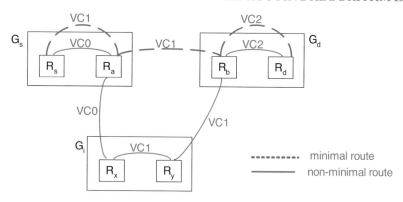

Figure 5.4: Routing and virtual channel assignment on the dragonfly topology.

AD uses a sequential allocator. The routing is identical to adaptive routing in a folded-Clos[40] where the folded-Clos is flattened into the routers of the flattened butterfly. Thus, the intermediate node is chosen from the closest common ancestors and not among all nodes. As a result, even though CLOS AD is non-minimal routing, the hop count is always equal or less than that of a corresponding folded-Clos network.

5.5.3   EXAMPLE 3: DRAGONFLY

The dragonfly topology routing differs as the routing consists of an intra-group and inter-group routing. Because the inter-group or the *global* channels are more expensive, the main objective is to load-balance these global channels with adaptive routing.

Minimal routing in a dragonfly from source node s attached to router R_s in group G_s to destination node d attached to router R_d in group G_d traverses a single global channel and is accomplished in three steps:

Step 1 : If $G_s \neq G_d$ and R_s does not have a connection to G_d, route within G_s from R_s to R_a, a router that has a global channel to G_d.

Step 2 : If $G_s \neq G_d$, traverse the global channel from R_a to reach router R_b in G_d.

Step 3 : If $R_b \neq R_d$, route within G_d from R_b to R_d.

Step 1 and Step 3 are intra-group routing while Step 2 is the inter-group routing. This minimal routing works well for load-balanced traffic, but results in very poor performance on adversarial traffic patterns.

To load-balance adversarial traffic patterns, Valiant's algorithm [66] can be applied at the system level — routing each packet first to a randomly-selected intermediate group G_i and then to its final destination d. Applying Valiant's algorithm to groups suffices to balance load on both the global and local channels. This randomized non-minimal routing traverses at most two global channels and requires five steps:

Step 1 : If $G_s \neq G_i$ and R_s does not have a connection to G_i, route within G_s from R_s to R_a, a router that has a global channel to G_i.

Step 2 : If $G_s \neq G_i$ traverse the global channel from R_a to reach router R_x in G_i.

Step 3 : If $G_i \neq G_d$ and R_x does not have a connection to G_d, route within G_i from R_x to R_y, a router that has a global channel to G_d.

Step 4 : If $G_i \neq G_d$, traverse the global channel from R_y to router R_b in G_d.

Step 5 : If $R_b \neq R_d$, route within G_d from R_b to R_d.

Figure 5.4 shows how VCs [21] are used to avoid routing deadlock. To prevent routing deadlock [19], two VCs are needed for minimal routing and three VCs are required for non-minimal routing. This assignment eliminates all channel dependencies due to routing. For some applications, additional virtual channels may be required to avoid protocol deadlock — e.g., for shared memory systems, separate sets of virtual channels are required for request and reply messages. Based on these descriptions of minimal and nonminimal routing, adaptive routing algorithms described earlier in this chapter can also be applied to the dragonfly topology.

SUMMARY

Although the topology determines the performance bounds, the routing algorithm is critical in determining how much of this performance can be realized. Recently proposed high-radix topologies rely on proper adaptive routing algorithms to load-balance both the minimal and non-minimal channels. High-radix networks also present interesting challenges to adaptive routing because of *indirectness* of network congestion information and we demonstrate how indirect adaptive routing is needed for these routing algorithms.

CHAPTER 6

Scalable Switch Microarchitecture

To enable high-radix topologies described in earlier chapters, a scalable switch microarchitecture is needed that can scale to a high port count. Conventional router microarchitecture for low-radix topologies had a limited number of ports (i.e., 6 to 8 ports) and thus, centralized arbitration could be used. However, arbitration logic is proportional the $O(k^2)$ where k is the router radix (number of input and output ports). In this chapter, we describe a baseline router design, similar to that used for a low-radix router [49, 55]. This design scales poorly to high radix due to the complexity of the allocators and the wiring needed to connect them to the input and output ports. To overcome this limitation while also providing high performance, we describe a hierarchical switch organization that uses intermediate buffering to decouple the allocation between inputs and outputs while reducing the amount of intermediate buffers required.

6.1 ROUTER MICROARCHITECTURE BASICS

A block diagram of the baseline router architecture is shown in Figure 6.1. Arriving data is stored in the input buffers. These input buffers are typically separated into several parallel virtual channels that can be used to prevent deadlock, implement priority classes, and increase throughput by allowing blocked packets to be passed. The input buffers and other router resources are allocated in fixed-size units called *flits*, and each packet is broken into one or more flits as shown in Figure 6.2(a).

The progression of a packet through this router can be separated into per-packet and per-flit steps. The per-packet actions are initiated as soon as the *header flit*, the first flit of a packet, arrives:

1. **Route computation (RC)** - based on information stored in the header, the output port of the packet is selected.

2. **Virtual-channel allocation (VA)** - a packet must gain exclusive access to a downstream virtual channel associated with the output port from route computation. Once these per-packet steps are completed, per-flit scheduling of the packet can begin.

3. **Switch allocation (SA)** - if there is a free buffer in its output virtual channel, a flit can vie for access to the crossbar.

4. **Switch traversal (ST)** - once a flit gains access to the crossbar, it can be transferred from its input buffers to its output and on to the downstream router.

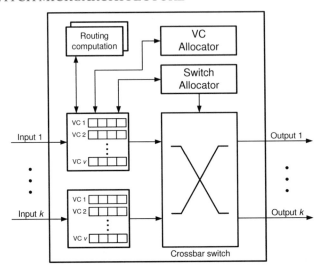

Figure 6.1: Baseline virtual channel router.

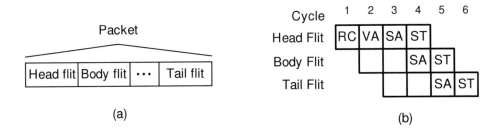

Figure 6.2: (a) Packets are broken into one or more flits (b) Example pipeline of flits through the baseline router.

These steps are repeated for each flit of the packet and upon the transmission of the *tail flit*, the final flit of a packet, the virtual channel is freed and is available for another packet. A simple pipeline diagram of this process is shown in Figure 6.2(b) for a three-flit packet assuming each step takes a single cycle.

6.2   SCALING BASELINE MICROARCHITECTURE TO HIGH RADIX

As radix is increased, a centralized approach to allocation rapidly becomes infeasible because the wiring, die area, and the latency all increase to prohibitive levels. In this section, we introduce distributed structures for both switch and virtual channel allocation that scale well to high port counts.

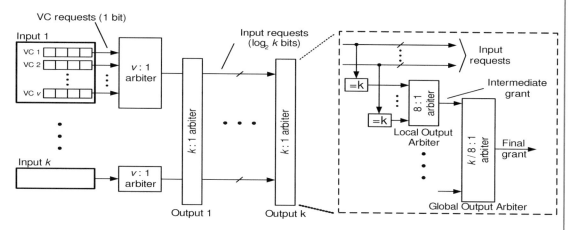

Figure 6.3: Scalable switch allocator architecture. The input arbiters are localized but the output arbiters are distributed across the router to limit wiring complexity. A detailed view of the output arbiter corresponding to output k is shown to the right.

We address the scalability of the switch allocator by using a distributed separable allocator design as shown in Figure 6.3. The allocation takes place in three stages: input arbitration, local output arbitration, and global output arbitration. During the first stage all ready virtual channels in each input controller request access to the crossbar switch. The winning virtual channel in each input controller then forwards its request to the appropriate local output arbiter by driving the binary code for the requested output onto a per-input set of horizontal request lines.

At each output arbiter, the input requests are decoded and, during stage two, each local output arbiter selects a request (if any) for its switch output from among a local group of m (in Figure 6.3, $m = 8$) input requests and forwards this request to the global output arbiter. Finally, the global output arbiter selects a request (if any) from among the k/m local output arbiters to be granted access to its switch output. For very high-radix routers, the two-stage output arbiter can be extended to a larger number of stages.

At each stage of the distributed arbiter, the arbitration decision is made over a relatively small number of inputs (typically 16 or less) such that each stage can fit in a clock cycle. For the first two stages, the arbitration is also local - selecting among requests that are physically co-located. For the final stage, the distributed request signals are collected via global wiring to allow the actual arbitration to be performed locally. Once the winning requester for an output is known, a grant signal is propagated back through to the requesting input virtual channel. To ensure fairness, the arbiter at each stage maintains a priority pointer which rotates in a round-robin manner based on the requests.

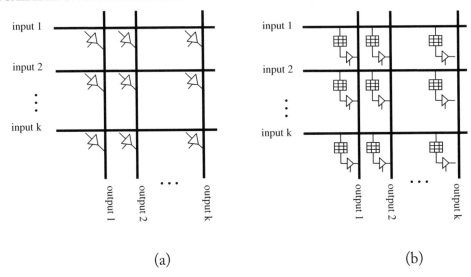

Figure 6.4: Block diagram of a (a) baseline crossbar switch and (b) fully buffered crossbar switch.

Virtual channel allocation (VA) poses an even more difficult problem than switch allocation because the number of resources to be allocated is multiplied by the number of virtual channels v. In contrast to switch allocation, where the availability of free downstream buffers is tracked with a credit count, with virtual channel allocation, the availability of downstream VCs is unknown. An ideal VC allocator would allow all input VCs to monitor the status of all output VCs they are waiting on. Such an allocator would be prohibitively expensive, with $v^2 k^2$ wiring complexity.

Building off the ideas developed for switch allocation, a scalable virtual channel allocator architectures can be built. The state of the output virtual channels are maintained at each crosspoint, and allocation is also performed at the crosspoints. However, VA involve *speculation* where switch allocation proceeds before virtual channel allocation is complete to reduce latency. Simple virtual channel speculation was proposed in [52] where the switch allocation and the VC allocation occurs in parallel to reduce the critical path through the router. With a deeper pipeline in a high-radix router, VC allocation is resolved later in the pipeline, which leads to more aggressive speculation

6.3   FULLY BUFFERED CROSSBAR

Adding buffering at the crosspoints of the switch (Figure 6.4b) decouples input and output virtual channel and switch allocation. This decoupling simplifies the allocation, reduces the need for speculation, and overcomes the performance problems of the baseline architecture with distributed, speculative allocators. Since input and output switch allocation are completely decoupled, a flit whose request wins the input arbitration is immediately forwarded to the crosspoint buffer corresponding to its output. At the crosspoint, local and global output arbitration are performed as in the unbuffered switch. However, because the flit is buffered at the crosspoint, it does not have to re-arbitrate at the input if it loses arbitration at the output.

The intermediate buffers are associated with the input VCs. In effect, the crosspoint buffers are per-output extensions of the input buffers. Thus, no VC allocation has to be performed to reach the crosspoint — the flit already holds the input VC. Output VC allocation is performed in two stages: a v-to-1 arbiter that selects a VC at each crosspoint followed by a k-to-1 arbiter that selects a crosspoint to communicate with the output.

To ensure that the crosspoint buffers never overflow, credit-based flow control is needed. Each input keeps a separate free buffer counter for each of the kv crosspoint buffers in its row. For each flit sent to one of these buffers, the corresponding free count is decremented. When a count is zero, no flit can be sent to the corresponding buffer. Likewise, when a flit departs a crosspoint buffer, a credit is returned to increment the input's free buffer count. The required size of the crosspoint buffers is determined by the credit latency – the latency between when the buffer count is decremented at the input and when the credit is returned in an unloaded switch.

It is possible for multiple crosspoints on the same input row to issue flits on the same cycle (to different outputs) and thus produce multiple credits in a single cycle. Communicating these credits back to the input efficiently presents a challenge. Dedicated credit wires from each crosspoint to the input would be prohibitively expensive. To avoid this cost, all crosspoints on a single input row share a single credit return bus. To return a credit, a crosspoint must arbitrate for access to this bus. The credit return bus arbiter is distributed, using the same local-global arbitration approach as the output switch arbiter.

With sufficient crosspoint buffers, this design achieves a saturation throughput of 100% of capacity because head-of-line blocking [36] is completely removed. As the amount of buffering at the crosspoints increases, the fully buffered architecture begins to resemble a virtual-output queued (VOQ) switch where each input maintains a separate buffer for each output. The advantage of the fully buffered crossbar compared to a VOQ switch is that there is no need for a complex allocator - the simple distributed allocation scheme discussed in Section 6.2 is able to achieve 100% throughput.

However, the performance benefits of a fully-buffered switch come at the cost of a much larger router area. The crosspoint buffering is proportional to vk^2 and dominates chip area as the radix increases. Figure 6.5 shows how storage and wire area grow with k in a $0.10\mu m$ technology for v=4. The storage area includes crosspoint and input buffers. The wire area includes area for the crossbar itself as well as all control signals for arbitration and credit return. As radix is increased, the bandwidth of the crossbar (and hence its area) is held constant. The increase in wire area with radix is due to increased control complexity. For a radix greater than 50, storage area exceeds wire area.

6.4   HIERARCHICAL CROSSBAR ARCHITECTURE

To overcome the high cost (area) associated with the fully buffered crossbar, a hierarchical switch architecture can significantly reduce the amount of intermediate buffers required [42]. A block diagram of the hierarchical crossbar is shown in Figure 6.6. The hierarchical crossbar divides the

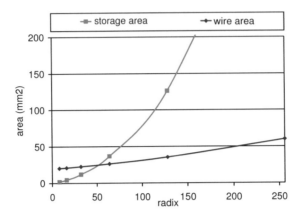

Figure 6.5: Area comparison between storage area and wire area in the fully buffered architecture.

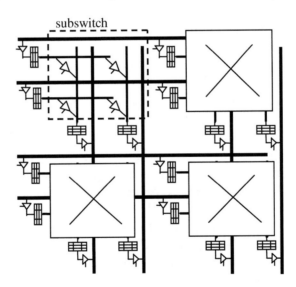

Figure 6.6: Hierarchical Crossbar ($k=4$) built from smaller subswitches ($p=2$).

crossbar switch into subswitches where only the inputs and outputs of the subswitch are buffered. A crossbar switch with k ports that has a subswitch of size p is made up of $(k/p)^2$ $p \times p$ crossbars, each with its own input and output buffers.

By implementing a subswitch design the total amount of buffer area grows as $O(vk^2/p)$, so by adjusting $p,$ the buffer area can be significantly reduced from the fully-buffered design. This architecture also provides a natural hierarchy in the control logic — local control logic only needs to consider information within a subswitch and global control logic coordinates the subswitches.

Similar to the fully-buffered architecture, the intermediate buffers on the subswitch boundaries are allocated on a per-VC basis. The subswitch input buffers are allocated according to a packet's *input* VC while the subswitch output buffers are allocated according to a packet's *output* VC. This decoupled allocation reduces HoL blocking when VC allocation fails and also eliminates the need to NACK flits in the intermediate buffers. By having this separation at the subswitches with buffers, it divides the VC allocation into a local VC allocation within the subswitch and a global VC allocation among the subswitches.

With the hierarchical design, an important design parameter is the size of the subswitch, p which can range from 1 to k. With small p, the switch resembles a fully-buffered crossbar resulting in high performance but also high cost. As p approaches the radix k, the switch resembles the baseline crossbar architecture giving low cost but also lower performance. In the next section, we describe the Cray YARC router [56] which implements this hierarchical organization with $k = 64$ and $p = 8$.

6.5   EXAMPLES OF HIGH-RADIX ROUTERS

With increasing pin bandwidth, we are seeing a paradigm shift to *many*-ported routers, along with *many-core* processors. As core count increases, the network ingress ports must also increase to avoid congestion and lock contention for shared resources at the sending host. This section describes two high-radix (k>32) routers, the Cray YARC and Mellanox InfiniScale IV. We focus on these because they provide raw bandwidth of 2.4Tb/s and 2.88Tb/s, respectively, yet have a fundamentally different microarchitecture.

6.5.1   CRAY YARC ROUTER

The Cray BlackWidow vector multiprocessor system [2], described in detail in Chapter 8, is one of the first systems to implement a high-radix network and YARC is the high-radix (radix-64) router used in the network that is based on the hierarchical organization described earlier in this chapter. The details of the YARC router can be found in [56], but in this section, we highlight some of the key differences between the YARC implementation and the hierarchical crossbar organization described earlier in Section 6.4.

A block diagram of the YARC router and a die photo is shown in Figure 6.7. The YARC router is a radix-64 router and the implementation is partitioned into 64 tiles with each tile containing an 8×8 subswitch, an input and an output port, and associated buffers which consist of input buffers, row buffers, and column buffers. The tiles communicate with other tiles through the row bus and the column channels. The tiled organization of the high-radix router led to a *complexity-effective* design as only a single design of a tile is required and is duplicated across the router. The die photo shown in Figure 6.7(b) shows the regular structure of the microarchitecture with a tile-based layout and the perimeter of the layout containing the SerDes (serializer/deserializer) I/O's.

The YARC implementation can be viewed as a two-stage network as shown in Figure 6.8 – the first stage consisting of the input speedup to the subswitches and the second stage consisting of output speedup to the output ports. Similar to a crossbar, there is only a single path between an input

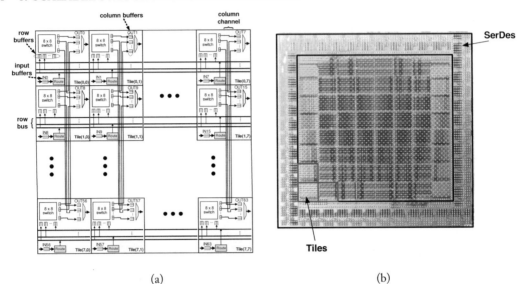

(a)   (b)

Figure 6.7: (a) Block diagram of the Cray YARC router and (b) die photo (courtesy Cray Inc).

and an output port but an $8\times$ speedup is provided at both the input and the output ports. Both the hierarchical organization (Section 6.4) and the YARC router provide an *input* speedup [20] since each input port is connected to all subswitches in its row. However, the YARC router exploits the abundant wire resources available on-chip as *output* speedup is also provided from the subswitches – i.e., the outputs of the subswitch are fully connected to all the outputs in each column. In comparison, a global bus was assumed for each output port in the hierarchical organization in Section 6.4. With the large number ports in a high-radix router, the output arbitration needs to be broken into multiple stages and the YARC router also performs output arbitration in two stages. The first stage arbitrates for the outputs of the subswitches and the second stage arbitrates for the output ports among the subswitches' outputs in each column. However, by providing output speedup, the output arbitration is simplified because the arbiter is local to the output port rather than being a central, shared resource.

Although there are abundant amount of wire resources available on-chip, the buffering available on-chip to implement the YARC router microarchitecture is limited. Thus, the intermediate buffers (row buffers and the column buffers) are area-constrained and the number of entries in these buffers are limited. As a result, although virtual cut-through flow control is implemented across YARC routers in the network, wormhole flow control is implemented within the YARC router – across row buffers and column buffers.

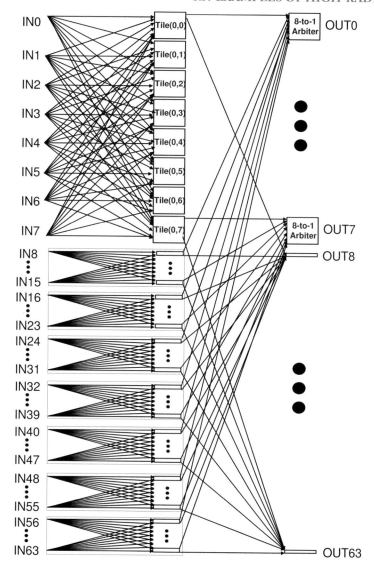

Figure 6.8: Block diagram of the Cray YARC router illustrating the internal speedup.

6.5.2 MELLANOX INFINISCALE IV

The Infiniband Trade Association (ITA) has a long-standing specification for a point-to-point IO communications. Over the years, it has evolved into a high-performance fabric through a combination of increased port count, and fast signaling speeds. The links in the InfinisScale IV (IS4) operate plesiochronously at data rates of 2.5Gb/s, 5Gb/s, and 10Gb/s. The *width* of the links can vary from 1× or 4× for a total link bandwidth of 10Gb/s (SDR), 20Gb/s (DDR) or 40Gb/s (QDR).

The microarchitecture of the IS4 takes a more conventional approach with a non-blocking 12×12 crossbar as the basic building block (Figure 6.9). The crossbars are replicated 3× to produce a 36-port router. Each host in the Infiniband fabric is labeled with a *local identifier* (LID)[1] Each crossbar uses a 48K entry *linear forwarding table* (LFT) to route unicast packets by indexing into the LFT using the *destination* LID.

(a) Packaged IS4 switch chip.

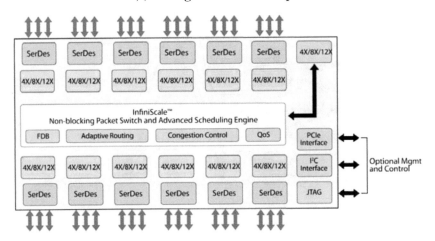

(b) Block diagram of the IS4 switch chip with 36 ports each 4×10 Gb/s, for an aggregate of 2.88 Tb/s off-chip bandwidth

Figure 6.9: Packaged silicon and block diagram of the Mellanox InfiniScale IV router.

[1]A LID is essentially the host endpoint or node identifier.

Several 4× ports can be aggregated to form a 8× or 12× port — providing 40, 80, or 120 Gb/s of bandwidth per direction, respectively. This allows the 36-ported 4× QDR router to be treated as a 12-port 12× QDR (120 Gb/s per direction) router which provides flexibility for building fat-trees, and torus networks with speedup in the network fabric, for example.

Each IS4 chip provide 16 *service levels* (SLs) with SL15 being reserved for control messages called *management datatgrams* (MADs). The SL is carried in the packet header and is invariant throughout the route. At each hop, a service level to *virtual lane* (VL) assignment takes place. The IS4 chip provides up to eight independent VLs which can be used for deadlock avoidance in the routing algorithm, performance isolation or QoS. The VLs use credit-based flow control to manage to downstream input buffer space and never will drop a packet due to congestion in the input buffer. Instead, the packet is blocked at the sender. If a different VL has room in the input buffer it may flow. Virtual cut-through flow control (VCT) [37] is used across the network links, with the exception of SL15 (the management SL) where *no* flow control is provided by the hardware. Software must provide the flow control in this case.

SUMMARY

As off-chip router bandwidth exponentially increases while typical packet sizes remain roughly constant, the increase in pin bandwidth relative to packet size motivates networks built from many *thin* links and create *high-radix* routers. However, the router microarchitecture needs to scale to a high port count effectively to enable a high-radix network. In this chapter, we described the challenges in scaling to high-radix – primarily the complexity of the switch and the virtual channel allocation that is proportional to the square of the radix. We presented an alternative hierarchical router microarchitectures and provided an example of a radix-64 Cray YARC router that leverages this hierarchical organization. By decoupling the input and the output allocation and reducing the intermediate buffering requirements, an hierarchical switch organization provides a cost-effective router microarchitecture that can scale to high port count.

CHAPTER 7

System Packaging

The packaging of various components imposes several constraints on the overall system design, such as topology, cooling, and cost. Since, ultimately, the system will be installed on the datacenter floor, the *packaging density* is the number of processing nodes per unit of area. The system must dissipate the heat which it generates, thus its *power density* describes the amount of power consumed per unit of area. The higher the power density, the more cooling required to dissipate the heat.

7.1 PACKAGING HIERARCHY

The system components, processing nodes and routers, are packaged within a *packaging hierarchy*. At the lowest level of the hierarchy are the compute modules which contain the processing nodes, and the routing modules which contain the switch chips. At the next level of the hierarchy, the modules may be connected via a backplane or midplane printed circuit board. Note, it may not be economical for a backplane because of airflow or cost limitations. The modules and backplane are contained within a *cabinet* or *rack* enclosure. The *system* consists of one or more rack enclosures with the necessary *cables* connecting the router ports according to the network topology. The network cables may aggregate multiple network links into a single cable to reduce both cost and cable bulk.

  For example, the Cray XT6 compute blade (Figure 7.1) densely packages eight processors along with their DRAM and network interface controllers (NICs) onto a single blade. A total of 24 blades are packaged into one *cabinet* (Figure 7.2) providing 192 multi-core processor sockets in a single cabinet[1]. A single cabinet provides 1536 processing cores, interconnected using either a 2-D or 3-D torus network. Blades are inserted from the front of the system (Figure 7.2a) into a backplane which aggregates multiple network links and brings them out to a connector on the back (Figure 7.2b).

7.2 POWER DELIVERY AND COOLING

The power delivery and cooling system must be designed to accommodate the *worst-case* power consumption at 100% utilization. In practice, however, a large cluster system rarely operates at full utilization. Nonetheless, with the cost of *operating* a large cluster largely determined by the energy cost [26] we want to deliver power from the utility to the datacenter as efficiently as possible. The power usage effectiveness (PUE) is the ratio of a datacenter's *total power* to the power *actually used* by computing equipment. According to a 2007 study by the United States Environmental Protection

---

[1]The two air-cooled cabinets shown in Figure 7.2 weigh about the same as a Volkswagen Beatle automobile!

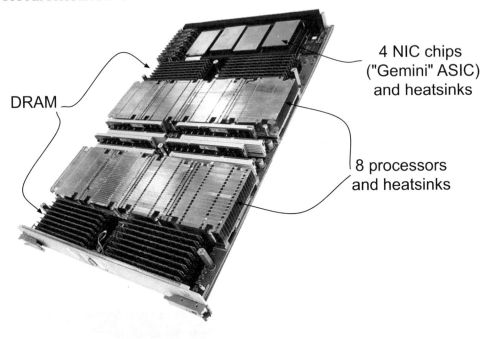

DRAM

4 NIC chips
("Gemini" ASIC)
and heatsinks

8 processors
and heatsinks

Figure 7.1: Cray XT6 compute blade with processors, DRAM, and NICs (source Cray Inc.)

Agency (EPA), the average datacenter PUE is 2.0 [65] and most efficient is 1.2 [30]. For example, assuming the average industrial electricity rate of $0.07 per kilowatt-hour (KWh) [64], each eight-processor Cray XT6 compute blade uses about 2KW of power resulting in an annual energy cost[2] in excess of $8M for a 32k processor system (Equation 7.1).

$$\frac{\$0.07}{\text{KWh}} \times \frac{24\,\text{h}}{\text{day}} \times \frac{2\,\text{KW}}{\text{blade}} \times 4096\,\text{blades} \times 1.6\,\text{PUE} = \$8.04\,\text{million} \qquad (7.1)$$

The *cooling* system must evacuate the heat generated by the processor sockets, DRAM, and networking equipment. Heat removal can be done via *convection* — blowing air across the hot components. Fans in each cabinet or rack are used to blow air across the component in combination with a *heat sink* to increase the surface area of the component, thereby improving its cooling efficiency. Copper or aluminum are common materials used for heat sinks. Although copper has twice the thermal conductivity as aluminum, it is also three times the weight. Figure 7.1 illustrates the use of copper as a heat sink for the processor and NIC chips.

---

[2]Equation 7.1 assumes a PUE of 1.6, which is the midpoint between the best-case (1.2) and average (2.0) PUE from the EPA's 2007 survey.

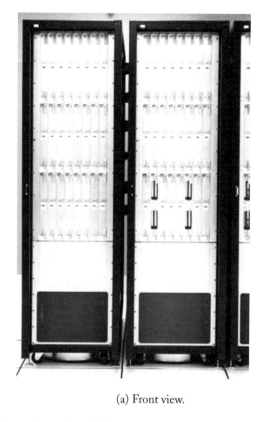

(a) Front view.   (b) Back view showing network cables.

Figure 7.2: Cray XT system packaging. (source Cray Inc.)

The exhaust air is then captured by an air return and passed through a cooling element (e.g., chiller or air conditioner) where the chilled air is recirculated and the process repeats. Direction of airflow is typically *front-to-back*, where cool air is drawn in from the front of the rack and exhausted out the back of the enclosure, or *back-to-front* with cool air enters from the front and warm air is exhausted out the front. The packaging constraints force a certain airflow, for example, if the cable bulk in the back of the machine impedes sufficient air intake. For this reason, the Cray XT (Figure 7.2) provides *bottom-to-top* airflow with a single large fan in the bottom of each cabinet. A 19-inch enclosure (Figure 7.3) is a standard rack for mounting common computer equipment such as servers and network switches in the datacenter. The pitch, or height, of each module in the rack is 1.75 inches — commonly referred to as one *rack unit*, or 1U for short. In practice, the actual pitch of the equipment that fits in a 1U slot allows about 30 *mils* (a mil is 1/1000 of an inch) of clearance to provide a gap for easier insertion and removal from the enclosure. An example of a Google rack is shown in Figure 7.4 circa 2003. Cabinets or racks are arranged in a sequence of *rows* and *columns* (or

(a) 19-inch rack.

(b) 1U module assembly.

Figure 7.3: An example a standard 19-inch rack common in many datacenter applications.

aisles) that maps the packaging onto the two-dimension floor surface of the datacenter, as shown in Figure 7.5. The airflow requirements in the datacenter often dictate that cabinets be arranged so that adjacent rows are back-to-back such that the exhausted air is dumped into a "hot aisle" and cool air drawn from a "cold aisle."

Another active cooling method is *water cooling*, which is less influenced by the ambient temperature in the datacenter. A water-cooled system uses plumbing in the cabinet or rack to circulate coolant through the system, as shown in Figure 7.6. Refrigeration units are distributed within the datacenter, as shown by the small black cabinets at the end of each aisle in Figure 7.5. Water is a common coolant used for such applications. Another agent commonly used to cool computer systems is *flourinert* [1], which is a non-toxic, non-flammable, and non-corrosive synthetic liquid. Flourinert can be synthesized to operate at a specific boiling point for single-phase liquid cooling applications, where it remains in liquid form. In a two-phase application, such as spray *evaporative cooling* where

Figure 7.4: An example Google rack with Xeon processors, circa 2003.

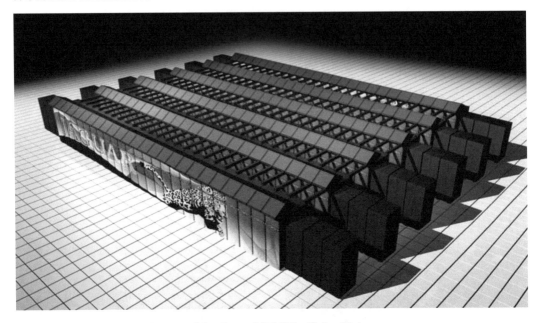

Courtesy of Cray Inc. and Oak Ridge National Laboratory

Figure 7.5: Example layout of a Cray XT system on the datacenter floor.

the Flourinert is injected directly on the surface of the die, where it boils, and undergoes a *phase change* from a liquid to a gas thereby removing the latent heat in the process. The process is similar to the cooling affects of perspiration and subsequent evaporation which the human body uses to cool itself.

The power distribution and cooling accounts for about 25% of the total datacenter cost [26]. The utility delivers power across transmission lines using 110KV (or above) to reduce energy loss across long distances. The incoming transmission lines are stepped down at a substation closer to end user, a datacenter in this case, to a 13KV line which is brought into an uninterruptible power supply (UPS) within the facility. The UPS is typically about 95% efficient, and is a small contributor to the *power usage effectiveness* (PUE) which is the ratio of a datacenter's total power to the power actually used by computing equipment. Efficient packaging, power, and cooling has a large impact on both the *capital* and *operating* cost of the cluster.

7.3   TOPOLOGY AND PACKAGING LOCALITY

One often overlooked property of a network is how a given topology maps to physical packaging. For example, a torus or mesh network which connect to their neighboring nodes makes most links

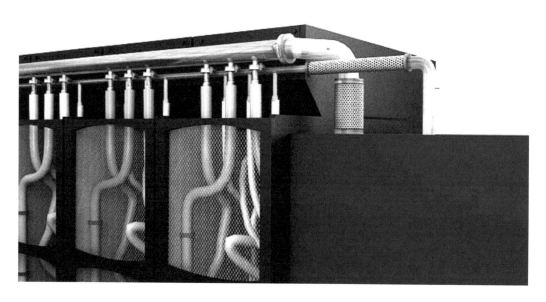

Figure 7.6: An example of plumbing in a liquid-cooled system (source Cray Inc.).

very short. The *wraparound* links in a torus can be made shorter by cabling the system as a *folded torus* as shown in (Figure 7.7). Mesh and torus networks have several packaging advantages:

(a) a portion of one dimension can be implemented on the printed circuit board (PCB) by connecting the adjacent nodes on the same board with PCB trace,

(b) a portion of one dimension can be implemented within a backplane PCB to connect adjacent nodes within the same rack or cabinet enclosure,

(c) cabling the mesh or torus is very regular,

(d) require relatively short cables which can operate at high signal rates and generally have a cost advantage over longer cables, and

(e) requires only a small number of different cable lengths.

Items (a) and (b) relate directly to *packaging locality* — nodes close together are connected together and can be *aggregated* since they originate and terminate near to one another. A direct network often has this quality of packaging locality. A folded-Clos, an indirect network, for example, has links from each router going to *different* terminating points since they connect to different switches in the next stage of the network. Item (c) is helpful from a manufacturing and deployment perspective, since a

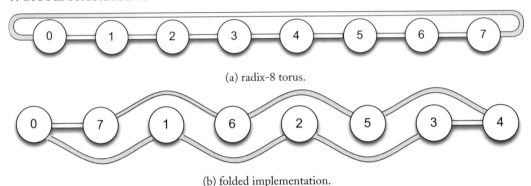

(a) radix-8 torus.

(b) folded implementation.

Figure 7.7: Decreasing the longest cable length in a torus (a) by "folding" it (b).

very regular cabling diagram is less difficult to correctly cable. Items (d) and (e) point out that torus and mesh networks have both shorter links, and just a few different cable lengths to interconnect nodes in different cabinets as shown in Figure 7.7b. Keeping the cables short[3] eliminates the need for repeaters or expensive optical links, and allows for high-speed serial point-to-point communication, with signal rates in excess of 10 Gb/s commonplace over a few meters.

The flattened butterfly is another example of a topology with a lot of packaging locality. A k-ary n-flat will have k switches co-located with each cabinet, where each switch has a minimum of p ports (Equation 7.2).

$$p \geq (n - 1)(k - 1) + k \qquad (7.2)$$

Each switch will use k electrical links to connect to its hosts, and another $\frac{1}{2}k - 1$ links to interconnect the *other* switches[4] (call this dimension 1) as shown in Figure 7.8. The total number of electrical links used (Equation 7.3)

$$e = k + (k - 1) \qquad (7.3)$$

Assume that all *inter-cabinet* links for the remaining $n - 2$ dimensions will require optics since it is likely they will be in excess of a few meters. Each switch connects $k - 1$ ports to cabinets in the *same row* (dimension 2), and another $k - 1$ ports to cabinets in the *same column* (dimension 3) as shown in Figure 7.8. More generally, the number of optical links in a k-ary n-flat is given by Equation 7.4.

$$o = (n - 2)(k - 1) \qquad (7.4)$$

The fraction of electrical links in the network is given by Equation 7.5. For example, an 8-ary 5-flat with 32k nodes, will use about 42% low cost, electrical links.

$$f_e = \frac{k + (k - 1)}{(n - 1)(k - 1) + k} \qquad (7.5)$$

[3]We consider "short" distances as cables shorter than 5m.
[4]These could be arranged as a stack of 1U switches co-located with the hosts, for instance.

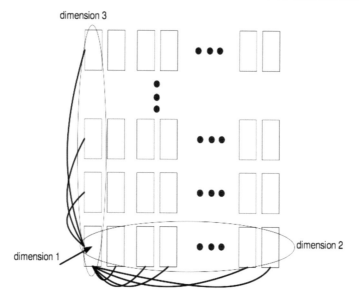

Figure 7.8: Example of how a flattened butterfly with four dimensions would map to a two-dimensional datacenter floor.

SUMMARY

The cost of power and its associated delivery and cooling are becoming significant factors in the cost of large datacenters. The interconnection network in a large parallel computer plays a central role both in its cost and performance. The way a system is packaged will ultimately influence the design of the network since it impacts the topology, cable reach, signaling technology, and cost per unit of bandwidth. A large-scale parallel computer is packaged with different levels of the packaging hierarchy, which must be efficiently mapped onto the two-dimensional floor of a datacenter.

CHAPTER 8

Case Studies

This Chapter is created to be a standalone entity; as such, it may repeat some of the concepts (e.g., flits, phits, routing, etc.) that have already been covered thus far. That is intentional. We want the reader to see how everything fits together and be able to look back at previous Chapters if questions arise.

8.1 CRAY BLACKWIDOW MULTIPROCESSOR

The Cray BlackWidow (BW) vector multiprocessor is designed to run demanding applications with high communication and memory bandwidth requirements. It uses a distributed shared memory (DSM) architecture to provide the programmer with the *appearance* of a large globally shared memory with direct load/store access. Unlike conventional microprocessors, each BW processor supports abundant memory level parallelism (MLP), with up to 4K outstanding global memory references per processor. Latency hiding and efficient synchronization are central to the BW design, and the network must therefore provide high global bandwidth while also providing low latency for efficient synchronization. The high-radix folded-Clos network [56] allows the system to scale up to 32K processors with a worst-case diameter of seven hops.

8.1.1 BLACKWIDOW NODE ORGANIZATION

Figure 8.1 shows a block diagram of a BlackWidow compute *node* consisting of four BW processors, and 16 Weaver chips with their associated DDR2 memory parts co-located on a *memory daughter card* (MDC). The processor to memory channels between each BW chip and Weaver chip use a 4-bit wide 6.25 Gbaud serializer/deserializer (SerDes) for an aggregate channel bandwidth of 16×3.125 Gbytes/s = 50 Gbytes/s per direction — 200 Gbytes/s per direction for each node.

The Weaver chips serve as pin expanders, converting a small number of high-speed differential signals from the BW processors into a large number of single-ended signals that interface to commodity DDR2 memory parts. Each Weaver chip manages four DDR2 memory channels, each with a 32-bit of data, 7-bit error correcting code (ECC), and one "spare bit". The 32-bit data path, coupled with the four-deep memory access bursts of DDR2, provides a minimum transfer granularity of only 16 bytes. Thus, the BlackWidow memory daughter card has twice the peak data bandwidth and four times the single-word bandwidth of a standard 72-bit-wide DIMM. Each of the eight MDCs contains 20 or 40 memory parts, providing up to 128 Gbytes of memory capacity per node using 1-Gbit memory parts.

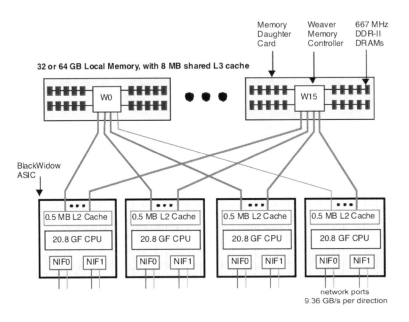

Figure 8.1: BlackWidow node organization.

8.1.2   HIGH-RADIX FOLDED-CLOS NETWORK

To reduce the cost and the latency of the network, BlackWidow uses a folded-Clos [14] network that is modified by adding *sidelinks* that connect peer subtrees and statically partition the global network bandwidth. Deterministic routing is performed using a hash function to obliviously balance network traffic while maintaining ordering on a cache line basis. Machines of up to 1024 processors can be constructed by connecting up to 32 rank 1 (R1) subtrees, each with 32 processors, to rank 2 (R2) routers. Machines of up to 4608 processors can be constructed by connecting up to nine 512-processor R2 subtrees via side links. Up to 16K processors may be connected by a rank 3 (R3) network where up to 32 512-processor R2 subtrees are connected by R3 routers. Multiple R3 subtrees can be interconnected using sidelinks to scale up to 32K processors.

The BlackWidow system topology and packaging scheme enables very flexible provisioning of network bandwidth. For instance, by only using a single rank 1 router module, instead of two as shown in Figure 8.1.2a, the port bandwidth of each processor is reduced in half — halving both the cost of the network and its global bandwidth. An additional bandwidth *taper* can be achieved by connecting only a subset of the rank 1 to rank 2 network cables, reducing cabling cost and R2 router cost at the expense of the bandwidth taper as shown by the $\frac{1}{4}$ taper in Figure 8.1.2b.

The network is built using a high-radix router, which provides 64 ports $\times 3$ lanes operating up to 6.25 Gb/s each lane. Each YARC router has an aggregate bandwidth of 2.4 Tb/s. BlackWidow scales up to 32K processors with a worst-case diameter of seven hops. YARC uses a hierarchical

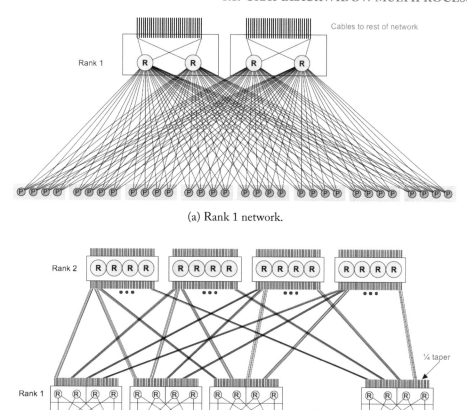

(a) Rank 1 network.

(b) Rank 2 network, shown with a $\frac{1}{4}$ taper.

Figure 8.2: The BlackWidow high-radix network.

organization [42] to overcome the quadratic scaling of conventional input-buffered routers. A two-level hierarchy is organized as an 8×8 array of tiles. This organization simplifies arbitration with a minimal loss in performance. The tiled organization also resulted in a modular design that could be implemented in a short period of time.

8.1.3   SYSTEM PACKAGING

Each compute module contains two compute nodes, as shown in Figure 8.1.3(a) providing a dense packaging solution with eight BW processors and 32 MDCs. At the next level of the hierarchy (see Figure 8.1.3 (b)), a set of eight compute modules and four router cards, each containing two YARC router chips, are connected via a midplane into a *chassis*. The router cards are mounted orthogonally

(a) BlackWidow compute module with two nodes.

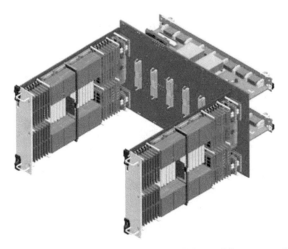

(b) BlackWidow chassis with eight compute modules and four network cards.

Figure 8.3: BlackWidow packaging.

to the compute blades, and each router chip connects to 32 of the 64 processors in the chassis. The chassis contains two rank-1 sub-trees, as shown in Figure 8.1.2(a).

All routing within a rank-1 sub-tree is carried via the PCB routing within the chassis. All routing between rank-1 sub-trees is carried over cables, which leave the back of the router cards. Two chassis are contained within one compute *cabinet* for a total of 128 BW processors providing an aggregate of ≈2.6 Tflops per cabinet. The BlackWidow system consists of one or more cabinets interconnected with the necessary cables using the high-radix folded-Clos [56] network.

8.1.4 HIGH-RADIX FAT-TREE

YARC is a high-radix router used in the network of the Cray BlackWidow multiprocessor. Using YARC routers, each with 64 3-bit wide ports, the BlackWidow scales up to 32K processors using a folded-Clos topology with a worst-case diameter of seven hops. Each YARC router has an aggregate bandwidth of 2.4Tb/s and a 32K-processor BlackWidow system has a bisection bandwidth of 2.5Pb/s.

YARC uses a hierarchical organization[42] to overcome the quadratic scaling of conventional input-buffered routers. A two-level hierarchy is organized as an 8×8 array of tiles. This organization simplifies arbitration with a minimal loss in performance. The tiled organization also resulted in a modular design that could be implemented in a short period of time.

The architecture of YARC is strongly influenced by the constraints of modern ASIC technology. YARC takes advantage of abundant on-chip wiring to provide separate column buses from each subswitch to each output port, greatly simplifying output arbitration. To operate using limited on-chip buffering, YARC uses wormhole flow control internally while using virtual-cut-through flow control over external channels.

To reduce the cost and the latency of the network, BlackWidow uses a folded-Clos network that is modified by adding *sidelinks* that connect peer subtrees and statically partition the global network bandwidth. We showed the benefits of high-radix Clos, compared to the previous torus networks, in terms of fault tolerance, bandwidth spreading, and simpler routing algorithm. Both adaptive and deterministic routing algorithms are implemented in the network to provide load-balancing across the network and still maintain ordering on memory requests. Deterministic routing is performed using a robust hash function to obliviously balance load while maintaining ordering on a cache line basis.

8.1.5   PACKET FORMAT

The format of a packet within the BlackWidow network is shown in Figure 8.4. Packets are divided into 24-bit phits for transmission over internal YARC datapaths. These phits are further serialized for transmission over 3-bit wide network channels. A minimum packet contains 4 phits carrying 32 payload bits. Longer packets are constructed by inserting additional payload phits (like the third phit in the figure) before the tail phit. Two-bits of each phit, as well as all of the tail phit are used by the data-link layer.

The head phit of the packet controls routing in addition to specifying the destination; this phit contains a v bit that specifies which virtual channel to use, and three bits, h, a, and r, that control specifically *how* the packet is routed. If the r bit is set, the packet will employ source routing. In this case, the packet header will be accompanied by a routing vector that indicates the path through the network as a list of ports to select the output port at each hop. Source routed packets are used only for maintenance operations such as reading and writing configuration registers on the YARC. If the a bit is set, the packet will route adaptively; otherwise, it will route deterministically. If the h bit is set, the deterministic routing algorithm employs the hash bits in the second phit to select the output port.

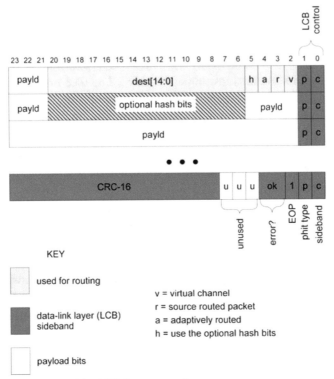

Figure 8.4: Packet format of the BlackWidow network.

8.1.6   NETWORK LAYER FLOW CONTROL

The allocation unit for flow control is a 24-bit phit — thus, the phit is really the flit (flow control unit). The BlackWidow network uses two virtual channels (VCs) [21], designated request (v=0) and response (v=1) to avoid request-response deadlocks in the network. Therefore, all buffer resources are allocated according to the virtual channel bit in the head phit. Each input buffer is 256 phits and is sized to cover the round-trip latency across the network channel. *Virtual cut-through* flow control [37] is used across the network links.

8.1.7   DATA-LINK LAYER PROTOCOL

The YARC data-link layer protocol is implemented by the link control block (LCB). The LCB receives phits from the router core and injects them into the serializer logic where they are transmitted over the physical medium. The primary function of the LCB is to reliably transmit packets over the network links using a sliding window go-back-N protocol. The send buffer storage and retry is on a packet granularity.

The 24-bit phit uses 2-bits of sideband dedicated as a control channel for the LCB to carry sequence numbers and status information. The virtual channel acknowledgment status bits travel in the LCB sideband. These VC acks are used to increment the per-vc credit counters in the output port logic. The *ok* field in the EOP phit indicates if the packet is healthy, encountered a transmission error on the current link (*transmit_error*), or was corrupted prior to transmission (*soft_error*). The YARC internal datapath uses the CRC to detect soft errors in the pipeline data paths and static memories used for storage. Before transmitting a tail phit onto the network link, the LCB will check the current CRC against the packet contents to determine if a soft error has corrupted the packet. If the packet is corrupted, it is marked as *soft_error*, and a good CRC is generated so that it is not detected by the receiver as a transmission error. The packet will continue to flow through the network marked as a bad packet with a soft error and eventually be discarded by the network interface at the destination processor.

The narrow links of a high-radix router cause a higher serialization latency to squeeze the packet over a link. For example, a 32B cache-line write results in a packet with 19 phits (6 header, 12 data, and 1 EOP). Consequently, the LCB passes phits up to the higher-level logic *speculatively*, prior to verifying the packet CRC, which avoids store-and-forward serialization latency at each hop. However, this early forwarding complicates various error conditions in order to correctly handle a packet with a transmission error and reclaim the space in the input queue at the receiver.

Because a packet with a transmission error is speculatively passed up to the router core and may have already flowed to the next router by the time the tail phit is processed, the LCB and input queue must prevent corrupting the router state. The LCB detects packet CRC errors and marks the packet as *transmit_error* with a corrected CRC before handing the end-of-packet (EOP) phit up to the router core. The LCB also monitors the packet length of the received data stream and *clips* any packets that exceed the *maximum packet length*, which is programmed into an LCB configuration register. When a packet is clipped, an EOP phit is appended to the truncated packet and it is marked as *transmit_error*. On either error, the LCB will enter error recovery mode and await the retransmission.

The input queue in the router must protect from overflow. If it receives more phits than can be stored, the input queue logic will adjust the tail pointer to excise the bad packet and discard further phits from the LCB until the EOP phit is received. If a packet marked *transmit_error* is received at the input buffer, we want to drop the packet and avoid sending any virtual channel acknowledgments. The sender will eventually timeout and retransmit the packet. If the bad packet has not yet flowed out of the input buffer, it can simply be removed by setting the tail pointer of the queue to the tail of the previous packet. Otherwise, if the packet has flowed out of the input buffer, we let the packet go and decrement the number of virtual channel acknowledgments to send by the size of the bad packet. The transmit-side router core does not need to know anything about recovering from bad packets. All effects of the error are contained within the LCB and YARC input queueing logic.

8.1.8   SERIALIZER/DESERIALIZER

The serializer/deserializer (SerDes) implements the *physical* layer of the communication stack. YARC instantiates a high-speed SerDes in which each *lane* consists of two complimentary signals making a balanced differential pair.

The SerDes is organized as a *macro* which replicates multiple lanes. For full duplex operation, we must instantiate the 8-lane receiver as well as an 8-lane transmitter macro. YARC instantiates 48 8-lane SerDes macros, 24 8-lane transmit and 24 8-lane receive macros, consuming ≈ 91.32 mm^2 of the 289 mm^2 die area, which is almost 1/3 of the available silicon (Figure 6.7).

The SerDes supports two full-speed data rates: 5 Gbps or 6.25 Gbps. Each SerDes macro is capable of supporting full, half, and quarter data rates using clock dividers in the PLL module. This allows the following supported data rates: 6.25, 5.0, 3.125, 2.5, 1.5625, and 1.25 Gbps. We expect to be able to drive a 6 meter, 26 gauge cable at the full data rate of 6.25 Gbps, allowing for adequate PCB foil at both ends.

Each port on YARC is three bits wide, for a total of 384 low voltage differential signals coming off each router, 192 transmit and 192 receive. Since the SerDes macro is 8 lanes wide and each YARC port is only 3 lanes wide, a naive assignment of tiles to SerDes would have 2 and 2/3 ports (8 lanes) for each SerDes macro. Consequently, we must aggregate three SerDes macros (24 lanes) to share across eight YARC tiles (also 24 lanes). This grouping of eight tiles is called an *octant* and imposes the constraint that each *octant* must operate at the same data rate.

The SerDes has a 16/20 bit parallel interface which is managed by the link control block (LCB). The *positive* and *negative* components of each differential signal pair can be arbitrarily swapped between the transmit/receive pair. In addition, each of the 3 lanes which comprise the LCB port can be permuted or "swizzled." The LCB determines which are the positive and negative differential pairs during channel initialization, as well as which lanes are "swizzled". This degree of freedom simplifies the board-level river routing of the channels and reduces the number of metal layers on a PCB for the router module.

8.2   CRAY XT MULTIPROCESSOR

The Cray XT4 system scales up to 32k nodes using a bidirectional three-dimensional torus interconnection network. Each *node* in the system consists of an AMD64 superscalar processor connected to a Cray Seastar chip [13] (Figure 8.5) which provides the processor-network interface, and 6-ported router for interconnecting the nodes. The system supports an efficient distributed memory message passing programming model. The underlying message transport is handled by the Portals [11] messaging interface.

The Cray XT interconnection network has several key features that set it apart from other networks:

• scales up to 32K network endpoints,

• high injection bandwidth using HypterTransport (HT) links directly to the network interface,

- reliable link-level packet delivery in hardware,

- multiple virtual channels for both deadlock avoidance and performance isolation, and

- age-based arbitration to provide fair access to network resources.

There are two types of nodes in the Cray XT system. Endpoints (nodes) in the system are either *compute* or *system and IO (SIO)* nodes. SIO nodes are where user's login to the system and compile/launch applications.

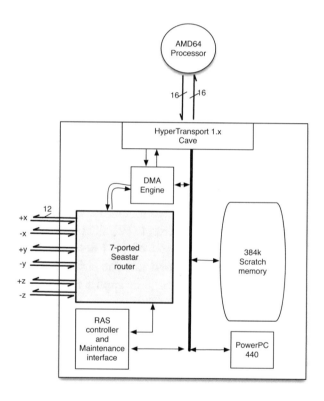

Figure 8.5: High level block diagram of the Seastar interconnect chip.

8.2.1   3-D TORUS

The Cray XT interconnect can be configured as either a *k*-ary *n*-mesh or *k*-ary *n*-cube (torus) topology. As a torus, the system is implemented as a *folded* torus to reduce the cable length of the wrap around link. The 7-ported Seastar router provides a processor port, and six network ports corresponding to +x, -x, +y, -y, +z, and -z directions. The port assignment for network links is not fixed, any port can correspond to any of the six directions. The non-coherent HyperTransport (HT) protocol provides a low latency, point-to-point channel used to drive the Seastar network interface.

Four virtual channels are used to provide point-to-point flow control and deadlock avoidance. Using virtual channels avoids unnecessary head-of-line (HoL) blocking for different network traffic flows, however, the extent to which virtual channels improve network utilization depends on the distribution of packets among the virtual channels.

8.2.2 ROUTING

The routing rules for the Cray XT are subject to several constraints. Foremost, the network must provide error-free transmission of each packet from the *source* node identifier (NID) to the *destination*. To accomplish this, the distributed *table-driven* routing algorithm is implemented with a dedicated *routing table* at each input port that is used to lookup the destination port and virtual channel of the incoming packet. The lookup table at each input port is not sized to cover the maximum 32k node network since most systems will be much smaller, only a few thousand nodes. Instead, a hierarchical routing scheme divides the node name space into *global* and *local* regions. The upper three bits of the destination field (given by the `destination[14:12]` in the packet header) of the incoming packet are compared to the global partition of the current SeaStar router. If the global partition does not match, then the packet is routed to the output port specified in the global lookup table (GLUT). The GLUT is indexed by `destination[14:12]` to choose one of eight global partitions.

Once the packet arrives at the correct global region, it will precisely route within a local partition of 4096 nodes given by the `destination[11:0]` field in the packet header. The tables must be constructed to avoid deadlocks. Glass and Ni [25] describe *turn* cycles that can occur in k-ary n-cube networks. However, torus networks are also susceptible to deadlock that results from overlapping virtual channel dependencies (this only applies to k-ary n-cubes, where $k > 4$) as described by Dally and Seitz [19]. Additionally, the SeaStar router does not allow 180 degree turns within the network. The routing algorithm must both provide deadlock-freedom and achieve good performance on benign traffic. In a fault-free network, a straightforward dimension-ordered routing (DOR) algorithm will provide balanced traffic across the network links. Although, in practice, faulty links will occur and the routing algorithm must route around the bad link in a way that preserves deadlock freedom and attempts to balance the load across the physical links. Furthermore, it is important to optimize the buffer space within the SeaStar router by balancing the number of packets within each virtual channel.

8.2.2.1 Avoiding deadlock in the presence of faults and turn constraints

The routing algorithm rests upon a set of rules to prevent deadlock. In the turn model, a positive first (x+, y+, z+ then x-, y-, z-) rule prevents deadlock and allows some routing options to avoid faulty links or nodes. The global/local routing table adds an additional constraint for valid turns. Packets must be able to travel to their local area of the destination without the deadlock rule preventing free movement within the local area. In the Cray XT network the localities are split with yz planes. To allow both x+ and x- movement without restricting later directions, the deadlock avoidance rule is modified to (x+, x-, y+, z+ then y+, y-, z+ then z+, z-). Thus, free movement is preserved. Note that missing or broken X links may induce a non-minimal route when a packet is routed via the global table (since only y+ and z+ are "safe"). With this rule, packets using the global table will prefer to move in the X direction, to get to their correct global region as quickly as possible. In the absence of any broken links, routes between compute nodes can be generated by moving in x dimension, then y, then z. Also, when y=Y_{max}, it is permissible to dodge y- then go x+/x-. If the dimension is configured as a mesh — there are no y+ links, for example, anywhere at y=Y_{max} then a deadlock cycle is not possible.

In the presence of a faulty link, the deadlock avoidance strategy depends on the direction prescribed by dimension order routing for a given destination. In addition, toroidal networks add *dateline* restrictions. Once a dateline is crossed in a given dimension, routing in a higher dimension (e.g., X is "higher" than Y) is not permitted.

8.2.2.2   Routing rules for X links

When x+ or x- is desired, but that link is broken, y+ is taken if available. This handles crossing from compute nodes to service nodes, where some X links are not present. If y+ is not available, z+ is taken. This z+ link must not cross a dateline. To avoid this, the dateline in Z is chosen so that there are no nodes with a broken X link and a broken y+ link. Although the desired X link is available, the routing algorithm may choose to take an alternate path when the node at the other side of the X link has a broken y+ and z+ link (note the y+ might not be present if configured as a mesh), then an early detour toward z+ is considered. If the X link crosses a partition boundary into the destination partition or the current partition matches the destination partition and the current Y matches the destination Y coordinate, route in z+ instead. Otherwise, the packet might be boxed in at the next node, with no safe way out.

8.2.2.3   Routing rules for Y links

When the desired route follows a Y link that is broken, the preference is to travel in z+ to find a good Y link. If z+ is also broken, it is feasible to travel in the opposite direction in the Y dimension. However, the routing in the node in that direction must now look ahead to avoid a 180 degree turn if it were to direct a packet to the node with the faulty links. When the desired Y link is available, it is necessary to check that the node at that next hop does not have a z+ link that the packet might prefer (based on XYZ routing) to follow next. That is, if the default direction for this destination in the next node is z+ and the z+ link is broken there, the routing choice at this node would be changed from the default Y link to z+.

8.2.2.4   Routing rules for Z links

When the desired route follows a z+ link that is broken, the preference is to travel in y+ to find a good z+ link. In this scenario, the Y link look ahead is relied up to avoid the node at y+ from sending the packet right back along y-. When the y+ link is not present (at the edge of the mesh), the second choice is y-. When the desired route is to travel in the z- direction, the logic must follow the z- path to ensure there are no broken links at all on the path to the final destination. If one is found, the route is forced to z+, effectively forcing the packet to go the long way around the Z torus.

8.2.3   FLOW CONTROL

Buffer resources are managed using credit-based flow control at the data-link level. The link control block (LCB) is shown at the periphery of the Seastar router chip in Figure 8.6. Packets flow across the network links using virtual cut-through flow control — that is, a packet does not start to flow until there is sufficient space in the receiving input buffer. Each virtual channel (VC) has dedicated buffer space. A 3-bit field (Figure 8.7) in each flit is used to designate the virtual channel, with a value of all 1's representing an *idle* flit. Idle flits are used to maintain byte and lane alignment across the plesiochronous channel. They can also carry VC credit information back to the sender.

8.2.4   SEASTAR ROUTER MICROARCHITECTURE

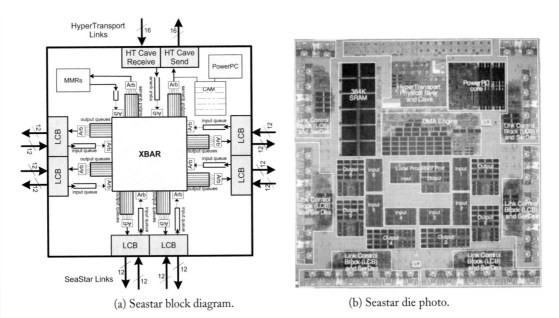

(a) Seastar block diagram.   (b) Seastar die photo.

Figure 8.6: Block diagram of the Seastar system chip.

Network packets are comprised of one or more 68-bit *flits* (flow control units). The first flit of the packet (Figure 8.7) is the *header* flit and contains all the necessary routing fields (destination[14:0], age[10:0], vc[2:0]) as well as a tail (t) bit to mark the end of a packet. Since most XT networks are on the order of several thousand nodes, the lookup table at each input port is not sized to cover the maximum 32k node network. To make the routing mechanism more space-efficient, the 15-bit node identifier is partitioned to allow a two-level hierarchical lookup: a small 8-entry table identifies a *region*, the second table precisely identifies the node within the region. The region table is indexed by the upper 3-bits of the *destination* field of the packet, and the low-order 12-bits identifies the node within 4k-entry table. Each network port has a dedicated routing table and is capable of routing a packet each cycle. This provides the necessary lookup bandwidth to route a new packet every cycle. However, if each input port used a 32k-entry lookup table, it would be sparsely populated for modest-sized systems, and use an extravagant amount of silicon area.

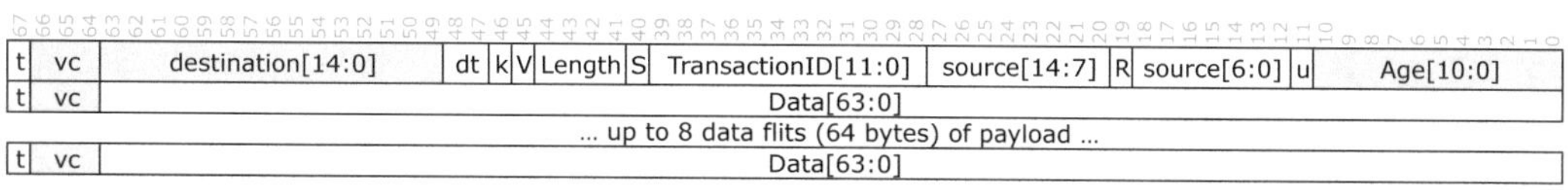

t	vc	destination[14:0]	dt	k	V	Length	S	TransactionID[11:0]	source[14:7]	R	source[6:0]	u	Age[10:0]
t	vc	Data[63:0]											
		... up to 8 data flits (64 bytes) of payload ...											
t	vc	Data[63:0]											

Figure 8.7: Seastar packet format.

A two-level hierarchical routing scheme is used to efficiently lookup the egress port at each router. Each router is assigned a unique node identifier, corresponding to its destination address. Upon arrival at the input port, the packet destination field is compared to the node identifier. If the upper three bits of the destination address match the upper three bits of the node identifier, then the packet is in the correct *global partition*. Otherwise, the upper three bits are used to index into the 8-entry *global lookup table* (GLUT) to determine the egress port. Conceptually, the 32k possible destinations are split into eight, 4k partitions denoted by bits destination[11:0] of the destination field.

The SeaStar router has six full-duplex network ports and one processor port that interfaces with the Tx/Rx DMA engine (Figure 8.6). The network channels operate at 3.2 Gb/s ×12 lanes over electrical wires, providing a peak of 4.8 GB/s per direction of network bandwidth. The link control block (LCB) implements a sliding window go-back-N link-layer protocol that provides reliable chip-to-chip communication over the network links. The router switch is both input-queued and output-queued. Each input port has four (one for each virtual channel) 96-entry buffers, with each entry storing one flit. The input buffer is sized to cover the round-trip latency across the network link at 3.2 Gb/s signal rates. There are 24 staging buffers in front of each output port, one for each input source (five network ports, and one processor port), each with four VCs. The staging buffers are only 16 entries deep and are sized to cover the crossbar arbitration round-trip latency. Virtual cut-through [37] flow control into the output staging buffers requires them to be at least 9 entries deep to cover the maximum packet size.

8.2.4.1 Age-based output arbitration

Packet latency is divided into two components: *queueing* and *router* latency. The total delay (T) of a packet through the network with H hops is the sum of the queueing and router delay.

$$T = HQ(\lambda) + Ht_r \tag{8.1}$$

where t_r is the per-hop router delay (which is ≈ 50 ns for the Seastar router). The queueing delay, $Q(\lambda)$, is a function of the offered load (λ) and described by the latency-bandwidth characteristics of the network. An approximation of $Q(\lambda)$ is given by an M/D/1 queue model (Figure 8.8).

$$Q(\lambda) = \frac{1}{1 - \lambda} \tag{8.2}$$

When there is very low offered load on the network, the $Q(\lambda)$ delay is negligible. However, as traffic intensity increases, and the network approaches saturation, the queueing delay will dominate the total packet latency.

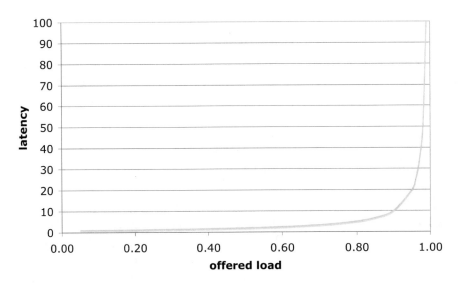

Figure 8.8: Offered load versus latency for an ideal M/D/1 queue model.

As traffic flows through the network it merges with newly injected packets and traffic from other directions in the network (Figure 8.9). This merging of traffic from different sources causes packets that have further to travel (more hops) to receive geometrically less bandwidth. For example, consider the 8-ary 1-mesh in Figure 8.9(a) where processors P0 thru P6 are sending to P7. The switch allocates the output port by granting packets fairly among the input ports. With a round-robin packet arbitration policy, the processor closest to the destination (P6 is only one hop away) will get the most bandwidth — 1/2 of the available bandwidth. The processor two hops away, P5, will get half of the bandwidth into router node 6, for a total of $1/2 \times 1/2 = 1/4$ of the available bandwidth. That is, every two arbitration cycles node 7 will deliver a packet from source P6, and every four arbitration cycles it will deliver a packet from source P5. A packet will merge with traffic from at most $2n$ other ports since each router has $2n$ network ports with $2n-1$ from *other* directions and one from the processor port. In the worst case, a packet traveling H hops and merging with traffic from $2n$ other input ports, will have a latency of:

$$T_{worst} = \frac{L}{(2n)^H} \tag{8.3}$$

where L is the length of the message (number of packets), and n is the number of dimensions. In this example, P0 and P1 each receive $1/64$ of the available bandwidth into node 7, a factor of 32 times less than that of P6. Reducing the variation in bandwidth is critical for application performance, particularly as applications are scaled to increasingly higher processor counts. Topologies with a lower diameter will reduce the impact of merging traffic. A torus is less affected than a mesh of the same radix (Figure 8.9a and 8.9b), for example, since it has a lower diameter. With dimension-order routing (DOR), once a packet starts flowing on a given dimension it stays on that dimension until it reaches the ordinate of its destination.

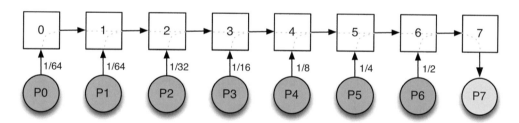

(a) 8-ary 1-dimensional mesh

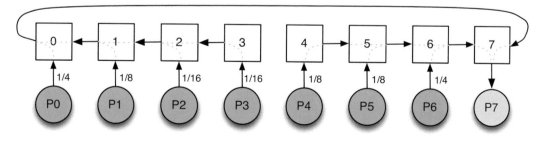

(b) 8-ary 1-dimensional torus

Figure 8.9: All nodes are sending to P7 and merging traffic at each hop.

8.2.4.2   Key parameters associated with age-based arbitration

The Cray XT network provides *age-based arbitration* to mitigate the affects of this traffic merging as shown in Figure 8.9, thus reducing the variation in packet delivery time. However, age-based arbitration can introduce a starvation scenario whereby *younger* packets are starved at the output port and cannot make forward progress toward the destination. The details of the algorithm along with performance results are given by Abts and Weisser [4]. There are three key parameters for controlling the aging algorithm.

- `AGE_CLOCK_PERIOD` – a chip-wide 32-bit countdown timer that controls the *rate* at which packets age. If the age rate is too slow, it will appear as though packets are not accruing any queueing delay, their ages will not change, and all packets will appear to have the same age. On the other hand, if the age rate is too fast, packets ages will saturate very quickly — perhaps after only a few hops — at the maximum age of 255, and packets will not generally be distinguishable by age. The resolution of `AGE_CLOCK_PERIOD` allows anywhere from 2 nanoseconds to more than 8 seconds of queueing delay to be accrued before the age value is incremented.

- `REQ_AGE_BIAS` and `RSP_AGE_BIAS` – each hop that a packet takes increments the packet age by the `REQ_AGE_BIAS` if the packet arrived on VC0/VC1 or by `RSP_AGE_BIAS` if the packet arrived on VC2/VC3. The age *bias* fields are configurable on a per-port basis, with the default bias of 1.

- `AGE_RR_SELECT` – a 64-bit array specifying the output arbitration policy. A value of all 0s will select *round-robin* arbitration, and a value of all 1s will select *age-based* arbitration. A combination of 0s and 1s will control the ratio of round-robin to age-based. For example, a value of $0101\cdots0101$ will use half round-robin and half age-based.

When a packet arrives at the head of the input queue, it undergoes routing by indexing into the LUT with destination[11:0] to choose the target port and virtual channel. Since each input port and VC has a dedicated buffer at the output staging buffer, there is no arbitration necessary to allocate the staging buffer — only flow control. At the output port, arbitration is performed on a per-packet basis (not per flit, as wormhole flow control would). Each output port is allocated by performing a 4-to-1 VC arbitration along with a 7-to-1 arbitration to select among the input ports. Each output port maintains two *independent* arbitration pointers — one for round-robin and one for age-based. A 6-bit counter is incremented on each *grant* cycle and indexes into the `AGE_RR_SELECT` bit array to choose the per-packet arbitration policy.

8.3   SUMMARY

The Cray BlackWidow is a scalable shared memory multiprocessor using custom vector processors, and the Cray XT is a distributed memory multiprocessor built from commodity microprocessors. The Cray XT uses a 3-D torus (low-radix) network, in contrast to the high-radix folded-Clos of the BlackWidow. This topology difference is in large part because the 3-D torus is a direct network and simply doesn't have silicon area to accommodate the additional SerDes. The BlackWidow network is an indirect network with the YARC switch chip having 192 SerDes surrounding the periphery of a 17x17mm die. The dense SerDes enabled a high-radix folded-Clos topology instead of a torus. More importantly, many scientific codes still have 3-D domain decomposition that exploits nearest neighbor communication and is best suited for a torus. So the topology choice is not only technology driven, but sometimes workload driven.

CHAPTER 9

Closing Remarks

Interconnection networks are the glue that binds the otherwise loosely-coupled distributed memory cluster systems that are common in datatcenter networks and the high-performance computing (HPC) community. The system scale — number of processor sockets capable of being housed in a single system — is impacted dramatically by the network. With *exa*-scale parallel computers being designed with 100s of thousands and even millions of processing cores, the cost of power and its associated delivery and cooling are becoming significant factors in the total expenditures of large-scale datacenters. Barroso and Hölzle recently showed a mismatch between common server workload profiles and server energy efficiency [8]. In particular, they show that a typical Google cluster spends most of its time within the 10-50% CPU utilization range, but that servers are inefficient at these levels. They therefore make the call for *energy proportional computing systems* that ideally consume almost no power when idle and gradually consume more power as the activity level increases. As of June, 2010 Top500 [62] list, the Cray XT5-HE with 224,162 processing cores achieving 1.759 petaflops and nearly 7 megawatts on the LINPACK benchmark[1].

9.1 PROGRAMMING MODELS

Warehouse-scale Computers (WSC) [9] such as those shown in Figure 1.1 fuel the Internet applications of today and tomorrow. WSC and HPC machines differ in programming models with datatcenter clusters dominated by TCP socket-based models, and distributed memory HPC systems commonly use message passing interfaces like (MPI), or hierarchical programming models that exploit shared memory (ccNUMA) within the node using an OpenMP interface and distributed memory between nodes with MPI. These differences result in $O(1\mu s)$ end-to-end message latency, compared to $O(100\mu s)$ of latency within datacenter servers. In large part, the software transport plays a critical role in latency — with TCP transport and multiple kernel-user space copies — confounding low-latency messaging. Efficient user-level messaging have been demonstrated with large-scale global communications on the order of $1\mu s$ in the HPC community, where efficient fine-grain communication and low-latency synchronization are hallmarks of scalable machines [7, 15, 35, 55].

9.2 WIRE PROTOCOLS

Supercomputers often take the design approach of building the entire machine from the most efficient packaging, chip technology, and signaling. As a result, they typically don't have a high-

---

[1]It is worth emphasizing that the Top500 list is simply a measure of how well a parallel computer solves systems of dense linear algebra, and suitability to other tasks may vary.

volume market, noting the original Cray-1 [54] in 1977 set a goal of delivering one system per quarter. The proprietary signaling and wire protocols (packet formats, etc.) have made traditional supercomputers incompatible with other vendors. However, the emergence of 40 gigabit and 100 gigabit Ethernet, coupled with a common processor-network interface like PCIe-Gen3 ×16 has the potential to convolve datacenters and supercomputing into "super datacenters" — what Barroso and Hölzle refer to as warehouse-scale computers [9].

9.3   OPPORTUNITIES

Going forward, highly scalable machines capable of a *exa*flop computation will require low-diameter scalable networks. Moreover, *reliability* and *scalability* are inseparable. System designers need to focus on building power-efficient systems without sacrificing reliability in the hardware, while application programmers will need to accept that servers are vulnerable to faults in components such as the processors, memory, network, disks, etc., including system software; hardware can and *does* fail, and programmers need to focus on *fault-aware* applications that can detect the loss of a component and still function. Power-efficiency and reliability are the two largest impediments to continued scaling. When programmers from both the supercomputing and Internet (datacenter) communities embrace these concepts, we will benefit from greater interoperability and converged programming models and best practices.

Bibliography

[1] 3MCorporation. http://www.3m.com/product/information/Fluorinert-Electronic-Liquid.html. 66

[2] Dennis Abts, Abdulla Bataineh, Steve Scott, Greg Faanes, James Schwarzmeier, Eric Lundberg, Tim Johnson, Mike Bye, and Gerald Schwoerer. The Cray BlackWidow: A Highly Scalable Vector Multiprocessor. In *Proceedings of the International Conference for High Performance Computing, Networking, Storage, and Analysis (SC'07)*, Reno, NV, November 2007. DOI: 10.1145/1362622.1362646 57

[3] Dennis Abts, Natalie D. Enright Jerger, John Kim, Dan Gibson, and Mikko H. Lipasti. Achieving predictable performance through better memory controller placement in many-core cmps. In *ISCA '09: Proceedings of the 36th annual international symposium on Computer architecture*, pages 451–461, 2009. DOI: 10.1145/1555754.1555810 41

[4] Dennis Abts and Deborah Weisser. Age-based packet arbitration in large-radix k-ary n-cubes. In *SC '07: Proceedings of the 2007 ACM/IEEE conference on Supercomputing*, pages 1–11, 2007. DOI: 10.1145/1362622.1362630 88

[5] A. Agarwal. Limits on Interconnection Network Performance. *IEEE Trans. Parallel Distrib. Syst.*, 2(4):398–412, 1991. DOI: 10.1109/71.97897 19, 25

[6] Jung Ho Ahn, Nathan Binkert, Al Davis, Moray McLaren, and Robert S. Schreiber. Hyperx: topology, routing, and packaging of efficient large-scale networks. In *SC '09: Proceedings of the Conference on High Performance Computing Networking, Storage and Analysis*, pages 1–11, New York, NY, USA, 2009. ACM. DOI: 10.1145/1654059.1654101 37, 43

[7] Baba Arimilli, Ravi Arimilli, Vincente Chung, Scott Clark, Wolfgang Denzel, Ben Drerup, Torsten Hoefler, Jody Joyner, Jerry Lewis, Jian Li, Nan Ni, and Ram Rajamony. The Cray T3E Network: Adaptive Routing in a High Performance 3D Torus. In *Hot Interconnects 18*, Aug. 2010. 91

[8] Luiz André Barroso and Urs Hölzle. The case for energy-proportional computing. *Computer*, 40(12):33–37, 2007. DOI: 10.1109/MC.2007.443 91

[9] Luiz André Barroso and Urs Hölzle. *The Datacenter as a Computer: An Introduction to Design of Warehouse-scale Machines*. 2009. 1, 9, 91, 92

[10] Laxmi N. Bhuyan and Dharma P. Agrawal. Generalized hypercube and hyperbus structures for a computer network. *IEEE Trans. Computers*, 33(4):323–333, 1984. DOI: 10.1109/TC.1984.1676437 34

[11] Ron Brightwell, Bill Lawry, Arthur B. MacCabe, and Rolf Riesen. Portals 3.0: Protocol building blocks for low overhead communication. In *IPDPS '02: Proceedings of the 16th International Parallel and Distributed Processing Symposium*, page 268, Washington, DC, USA, 2002. IEEE Computer Society. DOI: 10.1109/IPDPS.2002.1016564 80

[12] Ron Brightwell, Kevin T. Pedretti, Keith D. Underwood, and Trammell Hudson. Seastar interconnect: Balanced bandwidth for scalable performance. *IEEE Micro*, 26(3):41–57, 2006. DOI: 10.1109/MM.2006.65 19

[13] Ron Brightwell, Kevin T. Pedretti, Keith D. Underwood, and Trammell Hudson. Seastar interconnect: Balanced bandwidth for scalable performance. *IEEE Micro*, 26(3):41–57, 2006. DOI: 10.1109/MM.2006.65 80

[14] C Clos. A Study of Non-Blocking Switching Networks. *The Bell System technical Journal*, 32(2):406–424, March 1953. 31, 74

[15] Cray Inc. Cray xt5 http://www.cray.com/products/xt5. 91

[16] Cray X1. http://www.cray.com/products/systems/x1/. 19

[17] Cray XT3. http://www.cray.com/xt3. 19

[18] W. J. Dally. Performance Analysis of k-ary n-cube Interconnection Networks. *IEEE Transactions on Computers*, 39(6):775–785, 1990. DOI: 10.1109/12.53599 19, 25

[19] W. J. Dally and C. L. Seitz. Deadlock-free message routing in multiprocessor interconnection networks. *IEEE Trans. Comput.*, 36(5):547–553, 1987. DOI: 10.1109/TC.1987.1676939 50, 82

[20] W. J. Dally and B. Towles. *Principles and Practices of Interconnection Networks*. 2004. 7, 9, 20, 28, 30, 58

[21] William J. Dally. Virtual-channel Flow Control. *IEEE Transactions on Parallel and Distributed Systems*, 3(2):194–205, 1992. DOI: 10.1109/71.127260 8, 41, 48, 50, 78

[22] J. Duato, A. Robles, F. Silla, and R. Beivide. A comparison of router architectures for virtual cut-through and wormhole switching in a now environment. *Journal of Parallel and Distributed Computing*, 61(2):224 – 253, 2001. DOI: 10.1006/jpdc.2000.1679 9

[23] EMCORE Connects Cables. http://www.emcore.com/fiber_optics/emcoreconnects. 29

[24] Patrick Geoffray and Torsten Hoefler. Adaptive routing strategies for modern high performance networks. In *HOTI '08: Proceedings of the 2008 16th IEEE Symposium on High Performance Interconnects*, pages 165–172, Washington, DC, USA, 2008. IEEE Computer Society. DOI: 10.1109/HOTI.2008.21 45

[25] Christopher J. Glass and Lionel M. Ni. The turn model for adaptive routing. In *ISCA '92: Proceedings of the 19th annual international symposium on Computer architecture*, pages 278–287, 1992. DOI: 10.1145/139669.140384 82

[26] Albert Greenberg, James Hamilton, David A. Maltz, and Parveen Patel. The cost of a cloud: research problems in data center networks. *SIGCOMM Comput. Commun. Rev.*, 39(1):68–73, 2009. DOI: 10.1145/1496091.1496103 63, 68

[27] Mark D. Hill and Michael R. Marty. Amdahl's law in the multicore era. *Computer*, 41(7):33–38, 2008. DOI: 10.1109/MC.2008.209 3

[28] Mark Horowitz, Chih-Kong Ken Yang, and Stefanos Sidiropoulos. High-Speed Electrical Signaling: Overview and Limitations. *IEEE Micro*, 18(1):12–24, 1998. DOI: 10.1109/40.653013 13

[29] HyperTransport Consortium. http://www.hypertransport.org. 13

[30] Google Inc. Efficient computingâ step 2: efficient datacenters. http://www.google.com/corporate/green/datacenters/step2.html. 64

[31] InfiniBand Trade Association. http://www.infinibandta.org. 13

[32] Intel Corporation. http://www.intel.com/technology/quickpath. 13

[33] Natalie Enright Jerger, Dana Vantrease, and Mikko Lipasti. An evaluation of server consolidation workloads for multi-core designs. In *Proceedings of the 2007 IEEE 10th International Symposium on Workload Characterization*, IISWC '07, pages 47–56, Washington, DC, USA, 2007. IEEE Computer Society. DOI: 10.1109/IISWC.2007.4362180 6

[34] Nan Jiang, John Kim, and William J. Dally. Indirect adaptive routing on large scale interconnection networks. In *ISCA '09: Proceedings of the 36th annual international symposium on Computer architecture*, pages 220–231, 2009. DOI: 10.1145/1555754.1555783 43, 44

[35] IBM journal of Research and Development staff. Overview of the ibm blue gene/p project. *IBM J. Res. Dev.*, 52(1/2):199–220, 2008. DOI: 10.1147/rd.521.0199 19, 91

[36] M. J. Karol, M. G. Hluchyj, and S. P. Morgan. Input versus Output Queueing on a Space-division Packet Switch. *IEEE Transactions on Communications*, COM-35(12):1347–1356, 1987. DOI: 10.1109/TCOM.1987.1096719 55

[37] Parviz Kermani and Leonard Kleinrock. Virtual cut-through: A new computer communication switching technique. *Computer Networks*, 3:267–286, 1979. DOI: 10.1016/0376-5075(79)90032-1 61, 78, 85

[38] John Kim, Wiliam J. Dally, Steve Scott, and Dennis Abts. Technology-driven, highly-scalable dragonfly topology. In *ISCA '08: Proceedings of the 35th International Symposium on Computer Architecture*, pages 77–88, 2008. DOI: 10.1145/1394608.1382129 29, 35, 38, 44

[39] John Kim, William Dally, Steve Scott, and Dennis Abts. Cost-efficient dragonfly topology for large-scale systems. *IEEE Micro*, 29:33–40, 2009. DOI: 10.1109/MM.2009.5 43

[40] John Kim, William J. Dally, and Dennis Abts. Adaptive Routing in High-radix Clos Network. In *International Conference for High Performance Computing, Networking, Storage, and Analysis (SC'06)*, Tampa, FL, November 2006. DOI: 10.1145/1188455.1188552 45, 48, 49

[41] John Kim, William J. Dally, and Dennis Abts. Flattened butterfly: a cost-efficient topology for high-radix networks. In *ISCA '07: Proceedings of the 34th annual international symposium on Computer architecture*, pages 126–137, 2007. DOI: 10.1145/1250662.1250679 29, 34

[42] John Kim, William J. Dally, Brian Towles, and Amit K. Gupta. Microarchitecture of a high-radix router. In *ISCA '05: Proceedings of the 32nd Annual International Symposium on Computer Architecture*, pages 420–431, 2005. DOI: 10.1145/1080695.1070005 27, 55, 75, 77

[43] James Laudon and Daniel Lenoski. The SGI Origin: A ccNUMA Highly Scalable Server. In *Proc. of the 24th Annual Int'l Symp. on Computer Architecture*, pages 241–251, 1997. DOI: 10.1109/ISCA.1997.604692 4, 19

[44] Charles E. Leiserson. Fat-trees: universal networks for hardware-efficient supercomputing. *IEEE Trans. Comput.*, 34(10):892–901, 1985. 34

[45] Daniel Lenoski, James Laudon, Kourosh Gharachorloo, Wolf-Dietrich Weber, Anoop Gupta, John Hennessy, Mark Horowitz, and Monica S. Lam. The stanford dash multiprocessor. *Computer*, 25(3):63–79, 1992. DOI: 10.1109/2.121510 4

[46] Luxtera Blazar LUX5010. http://www.luxtera.com/ products_blazar.htm. 29

[47] Luxtera Inc. White Paper: Fiber will displace copper sooner than you think. Technical report, November 2005. 29

[48] Partho P. Mishra, Dheeraj Sanghi, and Satish K. Tripathi. Tcp flow control in lossy networks: analysis and enhancement. In *Proceedings of the IFIP TC6 Working Conference on Computer Networks, Architecture, and Applications. on Computer networks, architecture and applications*, pages 181–192, Amsterdam, The Netherlands, The Netherlands, 1993. Elsevier Science Publishers B. V. 7

[49] S. Mukherjee, P. Bannon, S. Lang, A. Spink, and D. Webb. The Alpha 21364 network architecture. In *Hot Chips 9*, pages 113–117, Stanford, CA, August 2001. DOI: 10.1109/40.988687 19, 28, 51

[50] Ted Nesson and S. Lennart Johnsson. Romm routing on mesh and torus networks. In *SPAA '95: Proceedings of the seventh annual ACM symposium on Parallel algorithms and architectures*, pages 275–287, New York, NY, USA, 1995. ACM. DOI: 10.1145/215399.215455 41

[51] PCI Express . http://en.wikipedia.org/wiki/PCI_Express. 13

[52] Li Shiuan Peh and William J. Dally. A Delay Model for Router Micro-architectures. *IEEE Micro*, 21(1):26–34, 2001. DOI: 10.1109/40.903059 54

[53] Duncan Roweth and Trevor Jones. QsNetIII an Adaptively Routed Network for High Performance Computing. *High-Performance Interconnects, Symposium on*, 0:157–164, 2008. DOI: 10.1109/HOTI.2008.31 45

[54] Richard M. Russell. The cray-1 computer system. *Commun. ACM*, 21(1):63–72, 1978. DOI: 10.1145/359327.359336 3, 5, 92

[55] S. Scott and G. Thorson. The Cray T3E Network: Adaptive Routing in a High Performance 3D Torus. In *Hot Interconnects 4*, Aug. 1996. 19, 28, 51, 91

[56] Steve Scott, Dennis Abts, John Kim, and William J. Dally. The blackwidow high-radix clos network. In *ISCA '06: Proceedings of the 33rd annual international symposium on Computer Architecture*, pages 16–28, 2006. DOI: 10.1109/ISCA.2006.40 19, 57, 73, 76

[57] Daeho Seo, Akif Ali, Won-Taek Lim, Nauman Rafique, and Mithuna Thottethodi. Near-optimal worst-case throughput routing for two-dimensional mesh networks. In *Proc. of the International Symposium on Computer Architecture (ISCA)*, pages 432–443, 2005. DOI: 10.1145/1080695.1070006 41

[58] Arjun Singh. *Load-Balanced Routing in Interconnection Networks*. PhD thesis, Stanford University, 2005. 41, 42, 48

[59] Arjun Singh, William J. Dally, Amit K. Gupta, and Brian Towles. GOAL: A load-balanced adaptive routing algorithm for torus networks. In *Proc. of the International Symposium on Computer Architecture (ISCA)*, pages 194–205, San Diego, CA, June 2003. DOI: 10.1109/ISCA.2003.1207000 43

[60] Arjun Singh, William J. Dally, Amit K. Gupta, and Brian Towles. Adaptive channel queue routing on k-ary n-cubes. In *SPAA '04: Proceedings of the sixteenth annual ACM symposium on Parallelism in algorithms and architectures*, pages 11–19, New York, NY, USA, 2004. ACM Press. DOI: 10.1145/1007912.1007915 43

[61] Thomas L. Sterling, John Salmon, Donald J. Becker, and Daniel F. Savarese. *How to build a Beowulf: a guide to the implementation and application of PC clusters.* MIT Press, Cambridge, MA, USA, 1999. 3

[62] Top500 Supercomputer Sites. http://www.top500.org/. 5, 9, 91

[63] Brian Towles and William J. Dally. Worst-case traffic for oblivious routing functions. In *SPAA '02: Proceedings of the fourteenth annual ACM symposium on Parallel algorithms and architectures*, pages 1–8, 2002. DOI: 10.1145/564870.564872 42

[64] U.S. Department of Energy. Average retail price of electricity. http://www.eia.doe.gov/cneaf/electricity/epm/table5_6_a.html. 64

[65] U.S. Environmental Protection Agency. Report to congress on server and datacenter energy efficiency. Public Law 109-431, August 2, 2007. 64

[66] L. G. Valiant. A scheme for fast parallel communication. *SIAM Journal on Computing*, 11(2):350–361, 1982. DOI: 10.1137/0211027 41, 42, 48, 49

[67] Hangsheng Wang, Li Shiuan Peh, and Sharad Malik. Power-driven Design of Router Microarchitectures in On-chip Networks. In *Proc. of the 36th Annual IEEE/ACM Int'l Symposium on Microarchitecture*, pages 105–116, 2003. DOI: 10.1109/MICRO.2003.1253187 29

Authors' Biographies

DENNIS ABTS

Dennis Abts is a Member of Technical Staff at Google, where he is involved in the system architecture and design of next-generation large-scale clusters. His research interests include scalable interconnection networks, parallel computer system design, and fault tolerant computing. Prior to joining Google, Dennis was a Sr. Principal Engineer and System Architect for Cray Inc. where he was principally involved with the architecture and design of several large-scale parallel computers over the span of his 10+ year tenure at Cray. Including, the Cray XT3 (Red Storm) and XT4, Cray X1, Cray BlackWidow (XT5), and next-generation systems sponsored by the DARPA HPCS initiative. Abts received his Ph.D. in computer science from University of Minnesota.

JOHN KIM

John Kim is an assistant professor in the Department of Computer Science at KAIST (Korea Advanced Institute of Science and Technology). He received a Ph.D. in Electrical Engineering from Stanford University and a B.S. and M.Eng. in Electrical Engineering from Cornell University. Prior to graduate school, he worked as a design engineer on the design of several microprocessors at Motorola and Intel. His research focuses on parallel architectures, interconnection networks, and datacenter architectures, and his research is funded by Microsoft Research Asia and Samsung.

Printed in the United States
by Baker & Taylor Publisher Services